RETROSPECTIVES
ANTIPATTERNS

RETROSPECTIVES ANTIPATTERNS

Aino Vonge Corry

✦✦Addison-Wesley

Boston • Columbus • New York • San Francisco • Amsterdam • Cape Town
Dubai • London • Madrid • Milan • Munich • Paris • Montreal • Toronto • Delhi • Mexico City
São Paulo • Sydney • Hong Kong • Seoul • Singapore • Taipei • Tokyo

For information about buying this title in bulk quantities, or for special sales opportunities (which may include electronic versions; custom cover designs; and content particular to your business, training goals, marketing focus, or branding interests), please contact our corporate sales department at corpsales@pearsoned.com or (800) 382-3419.

For government sales inquiries, please contact governmentsales@pearsoned.com.

For questions about sales outside the U.S., please contact intlcs@pearson.com.

Visit us on the Web: informit.com/aw

Library of Congress Control Number: 2020941957

ISBN-13: 978-0-13-682336-0
ISBN-10: 0-13-682336-X

1 2020

Publisher
Mark L. Taub

Executive Editor
Greg Doench

Assistant Editor
Menka Mehta

Managing Producer
Sandra Schroeder

Sr. Content Producer
Julie B. Nahil

Project Manager
Aswini Kumar/
codeMantra

Copy Editor
Carol Lallier

Indexer
Ken Johnson

Proofreader
Donna Mulder

Cover Designer
Chuti Prasertsith

Compositor
codeMantra

This book is dedicated to my best friend, Ellen Agerbo, who, based on her experience with me as we worked on our CS degrees, our common master's thesis, and part of the PhD program, never believed I would be able to write a book. To quote her: "You were always excellent at coming up with ideas, but writing has never been your strong suit."

CONTENTS

FOREWORD

In this book, Aino tells the story of how our relationship began. However, there's more to the story. There's a continually unfolding story about how our connection has continued, deepened, and matured through the ensuing years. Over time Aino has become more than someone I've mentored. She has become a highly respected colleague and friend.

Aino knows that, at times, I can feel overprotective about the practice of retrospectives facilitation. As an author and early proponent of team continuous learning and improvement, I want team meetings to deliver these valuable outcomes, every time. I encourage team leaders to invest in setting aside focused time for team retrospectives. I want teams (and their organizations) to receive ever-increasing benefits from their retrospective practice.

Unfortunately, I often hear stories of retrospectives-in-name-only, retrospectives held primarily to check the "retro box," or retrospectives limited to listing the answers to two or three questions, resulting in few if any actionable plans. The storytellers generally follow on with comments about teams that feel these meetings are a waste of time. I can't blame them. Of course they feel that way. Those meetings take up time without providing the team with the benefits promised. In my overprotective state, I want to deny the name. Whatever these meeting are, I refute the idea that they are my idea of retrospectives.

As a consequence, I'm eager to engage with colleagues who communicate the word about leading effective retrospectives. It helps me know that I'm not alone. These days, whenever I see a conference or other event program with Aino Corry presenting on a retrospectives-related topic, I'm thrilled. I know I can relax. Those audiences (as well as the teams she works more closely with) will receive valuable information from Aino about the path to team improvement. I am happy to recommend her trainings to anyone who asks.

That's why I'm so happy to recommend this book. In it, Aino has shared a robust, curated list of antipatterns and how to avoid them. (And they will be familiar to every seasoned facilitator. I'm intimately acquainted with most of them.) *And* she has shared so much more than tips and techniques. If you read this book carefully, you will find a gold mine—with precious nuggets including her personal experiences, effective facilitation resources, and pointers for extracting yourself and your team when you're stuck in an antipattern.

Pick up this book. Study these antipatterns. Identify the ones that show up most often for you. Then make a plan for your next retrospective to include Aino's alternative solutions and improve your consequences. You'll be glad you did.

Best wishes for your future retrospectives,

Diana Larsen
Coauthor, *Agile Retrospectives: Making Good Teams Great*
Cofounder and Chief Connector, Agile Fluency Project LLC

PREFACE

THIS BOOK IS FOR YOU

I have wanted to write this book for you for a long time. Actually, I have worked on it for so long that it has become a running joke between my family and friends.

I imagine you sitting on the sofa in the evening, frustrated with all the things that you experience as a retrospectives facilitator, and you want to know that someone shares your pain. If that is the case, this book is for you. I am about to tell you about all my mistakes and how I saw them repeating themselves to an extent that enabled me to write patterns about them.

Don't be discouraged by my describing them as antipatterns: each has a solution as part of the antipattern description as well. Most of the solutions entail planning things differently the next time around; for example, I might give you ideas on how to remember to explain the reason for an activity. But there are also some real-time solutions that can be applied spontaneously in response to events during the retrospective—for example, ideas on how to encourage people to talk when they are silent.

When reading this book, you may notice that most of the antipatterns I describe, in the setting of a retrospective, could be found in the wild among other types of meetings as well. I would claim that the solutions described in the antipatterns could hold for whatever type of meeting you are facilitating. Because you do have a facilitator for all your meetings, right? Jutta Eckstein, in *Retrospectives for Organizational Change* (2019), also made the supported claim that the structure of a retrospective can be used in more settings than the cyclic team check-in that my context describes. She describes how retrospectives can also help you implement an organizational change on a much larger scale than the Scrum team we often relate the concept of retrospectives to.

If you are very new to all this facilitation business, it might be a good idea to read my book, because then you will be aware of all the challenges you might run into. On the other hand, who are we kidding? We almost never learn something before we need it, so you will need to make mistakes of your own before you can start to appreciate this book. Reading it now might only give you a laugh: "Did she really *do* that?" "What on earth made her think it was a good idea to say that?" But a good laugh is important, so there still might be value in it for you.

How the Book Came to Life

At the beginning of this millennium, I was a co-organizer and program chair at the JAOO Conference (now known as GOTO) in Aarhus, Denmark, where Linda Rising was an invited speaker. Among many things, Rising is the coauthor, with Mary Lynn Manns, of *Fearless Change* (2005), and she introduced me to Norm Kerth's seminal book *Project Retrospectives: A Handbook for Team Reviews* (2001). I always enjoy listening to talks by Rising, but this one was special. The idea of retrospectives intrigued me, and I could see how they could be very useful to a lot of the teams at our clients' organizations as well as teams in our own company. Reading the book after the conference did not discourage me, and in 2007, I started facilitating a few retrospectives.

The next step in my journey was to attend a course about facilitating retrospectives with Diana Larsen as teacher. This really opened my eyes to the possibilities and the challenges. I was lucky enough to assist Diana in later courses in Denmark, and as a teacher you learn even more than as a student. Ever since, I have been fascinated by the thought of helping people and teams reflect and learn. I started facilitating more retrospectives, first for my colleagues in the IT industry and later in other companies and settings. In the last 10 years, I have facilitated hundreds of retrospectives in dozens of organizations and delivered talks on retrospectives at conferences, geek nights, and, frankly, wherever people didn't get a chance to run away.

I read many books and articles and watched numerous presentations, both online and at conferences. I learned, of course, a lot of activities, but I soon understood that being a good retrospective facilitator is not just about knowing which activities to use—there is much more to it. I spent some time learning about body language, and *The Definitive Book of Body Language* (Pease & Pease 2004) gave me many insights about the importance of eye contact, handshakes, and the effects of positioning yourself in relation to other people in different ways. I drew inspiration from numerous books that were not directly aimed at facilitating retrospectives; see the References section at the end of the book.

Prompted by my husband's reminder of my poor memory, I kept notes over the years on what I saw and heard while facilitating retrospectives. I noted in a little black book the techniques I tried, what worked, and what didn't work to help teams make forward progress and avoid getting stuck in a rut. If I improvised a way of redirecting a discussion that was going around in circles, I wrote a few sentences on it. When a group of developers needed a prod in the right direction to keep their meetings constructive and useful, I used the little black book to remind myself what I had already tried and whether it worked. If I invented or stole a silly game or an exercise that got people moving when their minds and bodies were weary from sitting and talking, I jotted it down. The retrospectives in these notebooks to date add up to 296 retrospectives facilitated for 68 different teams in 27 different companies, and I know that I did a lot before I started taking notes.

When I started using the notebook, it proved helpful not just for reminding me about what I had done that worked or did not work but also for following up on experiments a team may have decided on in previous retrospectives. I also wrote plans A and B for each retrospective so that I had all my plans in one place and could get inspiration from my earlier work. Some people might rely on their memory for past successes and blunders, but that is not an option for me. When I started facilitating online retrospectives, I used online tools and initially forgot to make use of the notebook. After some less-than-successful online retrospectives, I learned that the notebook was equally useful in an online retrospective because it gives me a quick overview and helps me remember.

This book, which I started writing in October 2013, is a distillation of my little black book—or rather, books. At least it's a distillation of the bad parts, because this is a book of antipatterns. These are the traps people have fallen into, the mistakes I've made, and my best tips for getting out of the traps and fixing the mistakes. Facilitating retrospectives is never the same twice; if it is, then that's an antipattern in itself. I never want to stop learning how to make meetings more useful and how to get teams to work better together. I also take great pleasure in showing skeptical software developers, who just want to be left alone to type in code, what they can gain from communicating with their colleagues for a small part of their working week.

Although the process of learning never ends and I am still learning, I want to share with a wider circle what I have learned so far. A lot of great material can be found in books and online about how to facilitate a good retrospective, but over the years, I have seen many people struggling with the same problems. That's why I decided this book should be structured as a collection of antipatterns. Retrospectives are not easy to facilitate and are easily ruined, or at least made less efficient, and I think the time is ripe for a book of antipatterns. But don't be put off by the *negativity* of antipatterns. Every antipattern contains a refactored solution that has worked for me and the teams I help.

Also, when you read the Refactored Solution sections in this book, remember to consider your own context before applying my proposal. As Diana Larsen used to say whenever I asked her what to do in a retrospective: "It depends," by which she means that it depends on the context.

PREREQUISITES

I assume that since you are reading this book, you are familiar with retrospectives and the role of the facilitator. If you need a refresher, you could spend some time reading about retrospectives on the Internet and, of course, read *Agile Retrospectives: Making Good Teams Great* (Larsen & Derby 2006).

I have added a lot of info boxes in the book to explain concepts that you might not know about or have forgotten. Feel free to skip them if you are already aware of what the different concepts cover.

WHAT IS A RETROSPECTIVE?

A retrospective is a chance for a team to reflect and learn from the past within a structured meeting. The main aim is to inspect the situation and adapt to the reality. *Inspect and adapt* is the core of any agile process and was first popularized with the Japanese word *kaizen* in *The Machine That Changed the World* (Womack, Jones & Roos 1990). To get a true inspection and be able to adapt to a situation, we need to create an atmosphere of trust in which people can share what they have experienced. The facilitator makes sure that every voice is heard in some way and that the team decides together what to spend time discussing or doing cause analysis on. The outcome of a retrospective is often a few experiments that the team can make in order to improve how they work. Or, as Larsen and Derby (2006) put it, retrospectives are about "making good teams great!" A retrospective is also a time to share with your team how you have experienced different events since the last retrospective and to gain a clear understanding of each other. As my late father used to say, "To understand everything is to forgive everything."

THE FIVE PHASES OF A RETROSPECTIVE

Among the various definitions of retrospectives, *Agile Retrospectives* (Larsen & Derby 2006) describes the life of a retrospective as five phases:

1. **Set the Stage:** The facilitator creates an atmosphere of trust, makes sure everyone's voice is heard, looks at earlier experiments, defines the theme for the retrospective, and manages any other tasks necessary to start the retrospective.

2. **Gather Data:** The team gathers data (on experiences, events, tests, sales, etc.) for the time that the retrospective is focused on.

3. **Generate Insights:** The team looks behind the data to find the stories and the causes behind them. This phase can be done as a free discussion or a cause analysis.

4. **Decide What to Do:** The team decides together what experiments to carry out to improve the way they work together.

5. **Close the Retrospective:** The team decides who is responsible for following up on the experiments. The facilitator wraps up what happened and perhaps provides an evaluation of the retrospective if he or she feels it would be valuable—retrospective over the retrospective.

Often, I hear people claim that for short sprint retrospectives, it does not make sense to go through all five phases. But that way of thinking is exactly what leads to premature decision making, as described in depth later in Chapter 1, **Wheel of Fortune**.

WHAT IS A PATTERN?

A pattern is an abstract solution to an often recurring problem. Patterns are a means to disseminate experience in a literary form. Their names make up a common vocabulary for design, programming, or whatever domain for which the patterns describe solutions. A pattern can be a way to describe how things are done in an organization. A pattern contains a description of the context and the forces that define the problem you need to solve, the pattern solution, and the benefits and consequences of applying the pattern. A pattern also often refers to other patterns, because the consequences could be helped with a solution found in another pattern.

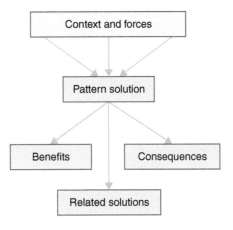

Figure P.1 The elements of a pattern

The concept of patterns was originated by the building architect Christopher Alexander and his coauthors in *A Pattern Language: Towns, Buildings, Construction* (1977). Just over a decade later, patterns were introduced for use in software by Kent Beck in a *Smalltalk Report* article, "A Short Introduction to Pattern Language" ((1993) 1999), with a focus on communication. Two years later, the concept was made popular by *Design Patterns: Elements of Reusable Object-Oriented Software* (Gamma et al. 1995), now known as the GoF book because the authors became collectively known as the Gang of Four.

When working with patterns, such as the Observer, Composite, and Strategy (Gamma et al. 1995), it is useful to refer to the names of patterns instead of having to explain a design or concept from scratch. Patterns are effective because of the way the brain works with cognitive patterns and cognitive automation. When you learn something, the details of that new knowledge are "glued" together in your long-term memory as a cognitive pattern. Together with the cognitive pattern, which helps you recognize the situation as a situation in which you have learned how to act, the cognitive automations, or how to react to the pattern, are also drilled into the brain. My former manager Michael Caspersen, who taught me almost everything I know about learning, includes many examples of and references to cognitive patterns and cognitive automations of the brain in his PhD dissertation (Caspersen 2007).

My PhD dissertation (Cornils 2001)[1] focused on software patterns, and I have noticed that I always see patterns in things, which is not uncommon among humans.

What Is an Antipattern?

An antipattern is a way to describe experience. The antipattern as a concept was first named and described by William J. Brown and his coauthors in *AntiPatterns: Refactoring Software, Architectures, and Projects in Crisis* (1998). It is a description of a solution to a frequently occurring problem in which the consequences outweigh the benefits.

The antipatterns in this book are the result of a facilitator not knowing better or not having the time or opportunity to do the right thing. Maybe the solution worked once for the facilitator in another group because the group members had a different way of communicating or knew each other better, but then it unexpectedly did not work in a new context.

I set the scene of antipatterns by giving you two examples of antipatterns. The first is an old one that led to a famous disaster. Late in the evening of April 14, 1912, the RMS *Titanic* hit an iceberg, and in the early hours of April 15, she sank, killing more than 1,500 of the 2,224 people on board, both passengers and crew. To understand why the ship sank, you have to look at a number of small things that, together, led to a disaster. The thing I choose to focus on is the antipattern you could call "following orders from an uninformed superior." If you look up *superior orders* on

---

1. Observant readers will realize that the name on my dissertation is not the name on this book. It is the name of my first husband, which was also my name at the time. Incidentally, this is also an antipattern: Do Not Change Your Name. At least not after publishing.

Wikipedia, you find, "Superior orders, often known as the Nuremberg defense, lawful orders, just following orders, or by the German phrase *Befehl ist Befehl* ('an order is an order'), is a plea in a court of law that a person—whether a member of the military, law enforcement, a firefighting force, or the civilian population—not be held guilty for actions ordered by a superior officer or an official." This antipattern has been identified in numerous places and times, and also, as it happens, in the middle of the tale of the *Titanic*.

The two wireless radio operators on the *Titanic* worked for the Marconi Wireless Telegraph Company. The operators' orders were to relay passengers' messages to and from friends and family on land in order to demonstrate the wireless communication service provided by the Marconi Company. It was also a factor that the company was paid for every message to and from the passengers, and thus their income increased by prioritizing them over the ship-to-ship courtesy messages. Almost from the beginning of the voyage, they had received warnings about icebergs and passed most of these messages to the bridge. Unfortunately, some of the messages sent to the *Titanic* were lost because the radio operators focused on following orders from their company. Their superior was the owner of the wireless company, not the captain.

This explains why, when at 9:40 p.m. the *Mesaba*, a ship sailing in the same waters as the *Titanic*, sent a warning of an ice field, the message never reached the bridge. At 10:55 p.m., another nearby cruise liner, the *Californian*, messaged that she had stopped after becoming surrounded by ice, but one of the radio operators on the *Titanic* scolded the *Californian* for interrupting him, since he was busy handling passenger messages. Consequently, the captain was not warned about the ice situation being worse than expected, and thus he continued to sail at full speed. It was not until 11:40 p.m., when an iceberg was spotted from the crow's nest, that the ship altered its route. The bridge crew started to turn the *Titanic*, but since she was a large ship sailing at high speed, it was too late. The side of the *Titanic* scraped along the iceberg, and the ship ruptured. We know how the story ends.

Often, a pattern in one context can be an antipattern in a different context. In the case of the *Titanic*, for example, the wireless radio operators were obligated to follow the orders of their superiors. However, had they known that the context had changed, that the ship faced an emergency, they would not have followed orders blindly—to do so would have constituted an antipattern.

Patterns also can become antipatterns over time, as technologies and processes change and improve. When a good solution is replaced by a better solution, the original solution can come to be viewed as a bad solution to a recurring problem.

The second example is the Microservices pattern, which is a design described by Martin Fowler and James Lewis (2014) whereby developers create a set of small services, each with its own functionality. The Microservices pattern proved to be maintainable, flexible, and resilient in software architectures, and it was hailed as the best thing since sliced bread (and design patterns). This pattern promoted the development of independently deployable, reusable components, enabling developers to create scalable systems built with microservices. The success of these systems led to the conversion of many monolithic systems to microservices architectures.

What happened then was what sometimes happens with patterns: the pattern was overused.[2] Microservices can have negative consequences if the organization lacks the expertise required to maintain the system, a domain that would benefit from this implementation, or well-defined boundaries between the services. Increased complexity is a major consequence of a microservices architecture, and in the wrong circumstances, the system becomes a more complex monolith instead of a suite of smaller, well-defined microservices. All the expected benefits, such as scalability, independence, and reusability, are lost, and if used in the wrong context, the Microservices pattern becomes an antipattern. The context in

---

2. The Singleton pattern from the GoF book is a brilliant example of pattern overuse.

which you use a pattern is important, and there will always be a context in which the pattern solution does not apply, resulting in an antipattern.

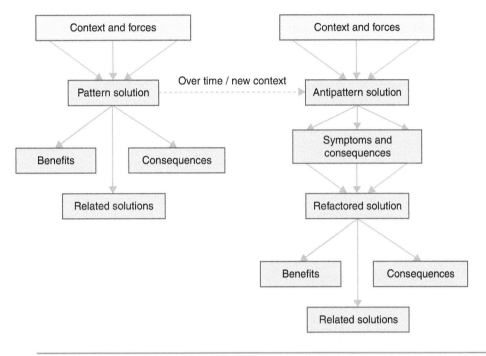

Figure P.2 A pattern becomes an antipattern when applied in the wrong context

An antipattern described correctly contains a general description, a list of the factors that led to the symptoms and how to recognize them, the consequences of the original solution, and a refactored solution that describes how to solve the current problems or at least how to do better next time.

All patterns have consequences. In some situations, it is a good idea to use a particular pattern, and in others, the same pattern becomes an antipattern. It is important to understand the context-dependent implications when you want to use a pattern. This helps you get the full picture, including the side effects of the antipattern solution. Like a pattern, an antipattern is not some abstract theory a person has invented but a series of causes and effects that he or she sees in often-recurring bad solutions.

You should read this book to learn to recognize antipatterns within retrospectives as (or perhaps even before) they happen to you. My goal in writing *Retrospectives Antipatterns* is to help you avoid making the same mistakes I have made so many times.

As an experienced retrospective facilitator, you may notice that you already know a lot of these patterns and how to deal with them. The added benefit of this book is that now you have a vocabulary to discuss it with other people, and it might be easier for you to recognize when you find yourself in an antipattern. If you share them with your colleagues, their memorable names can help to make you aware when you or your team is slipping into a **Wheel of Fortune** or **Prime Directive Ignorance**.

Lastly, you could read this book for the *schadenfreude*[3] because, as one of the authors of the original antipattern book, *Antipatterns: Refactoring Software, Architectures, and Projects in Crisis* (Brown et al. 1998), said at a presentation when that book was published, "[One's own] happiness is good, but the misfortune of others is better."

How to Read This Book

The Octopus

You might wonder what an octopus has to do with retrospectives. The short answer is: nothing. But that is just the short answer.

I got intrigued by octopuses when I learned about their intelligence. They can learn tricks by watching other octopuses being rewarded for learning tricks without even getting a reward themselves. They can crawl out of an aquarium and through a tiny pipe that leads to the ocean. They can climb out of an aquarium, get down on the floor, cross the floor, mount another table, enter another aquarium, eat all the fish there, and then go back to their own aquarium *as if nothing happened*.

---

3. *Schadenfreude* is a German word meaning that you take pleasure in other peoples' pain.

Most of all, I am stunned that 60 percent of an octopus's brain is in its eight legs, divided into eight individual little animals, almost, each with its own will and yet all working as a whole with the rest of the octopus in a synergy that is unique in the animal kingdom. I see a team with a facilitator at a retrospective as an octopus: the team and facilitator work together toward a common goal while still being individuals, with individual focus and strengths.

For every antipattern, there is an illustration with an octopus that captures the essence of the antipattern. In the antipattern **In the Soup**, the team works together to lift a weight that is still too heavy, because the problem they are trying to solve is **In the Soup** of things they cannot change but just need to accept. In **Prime Directive Ignorance**, the team starts blaming one person instead of trying to find the faults in the system. In the **Disillusioned Facilitator**, the team mocks the facilitator for trying out activities they find ridiculous.

THE LITERARY FORM OF ANTIPATTERNS IN THIS BOOK

Based on the literary form for antipatterns found in *Antipatterns* (Brown et al. 1998), I have decided on a specific form to make it easier to read the antipatterns. As you will see, this form fits better with some antipatterns than with others. For example, sometimes the symptoms are obvious, and other times they are subtle and worth describing in detail. For brevity, I have left out the section on forces that you would normally see in a pattern. In my literary form, I have folded the forces into the context and related context descriptions. Forces could be, for example, haste, eagerness to be heard, or lack of people.

Name: The names of patterns are important because names allow you to extend your vocabulary about retrospectives and enable you to communicate efficiently with others about those patterns. An interesting thing about giving something a name is that the name can encompass a large set of concepts, processes, and conditions in a succinct way that organizes information so that your brain can quickly access all of the elements associated with that name.

Context: This book is written as the learning journey of a retrospective facilitator from a Danish company. The story and the people will be introduced later, but for now it suffices to know that every antipattern will include a description of the context in which we find the antipattern in our story.

General Context: This section covers the environment you will be in if this antipattern occurs, a description of what may have led to this problem, and the urge to go for the antipattern solution. This is a more generic way of describing the situation in which you may find yourself tempted to implement the antipattern solution.

Antipattern Solution: This section explains the chosen path based on the problem described. The antipattern looked like a solution at the time, based on education, earlier experiences, time constraints, lack of courage, or simply orders from above. It might have been the right solution if it were not for the negative consequences in this context. If you are aware that you can end in this situation by choosing an antipattern solution, you might avoid making some of the numerous mistakes I have made. The antipattern solution is not to be confused with the refactored solution.

Consequences: All solutions and decisions have consequences. In the original book about design patterns (Gamma et al. 1995), they were listed as positive and negative consequences, and depending on context, one might outweigh the other. In antipatterns, the point is that the antipattern solution might fit well in another context, but in this context, the negative consequences are much larger than the benefits. I usually say that this listing of consequences is what makes patterns differ from mere methods or recipes found in other books.

Symptoms: Symptoms are the indicators enabling you to see that a particular retrospectives antipattern is occurring. Symptoms might include comments you hear either outside of or during the retrospective, behaviors you observe, moods you sense, and so on.

Refactored Solution: The refactored solution suggests how to improve the current situation so that you and your team are gaining more benefits than negative consequences. In some of the antipatterns, you will learn that it is not possible to implement the refactored solution once the situation has already arisen, and what you gain from such an antipattern is merely an awareness of how to avoid it next time.

Online Aspect: For some of the refactored solutions, there are differences in how you would overcome the challenges in online and offline settings. A growing number of the retrospectives I facilitate are online, so I share my experiences with this medium as well.

Personal Anecdote: In this section, I tell of my own experience in sighting the antipattern. Sometimes I found a way to refactor it while it happened, and sometimes it was a lesson learned for the next time.

OUTLINE OF THE ANTIPATTERNS

Structural Antipatterns

Structural antipatterns describe problems with the structure of the retrospective, such as how the activities are chosen, how the flow of the communication is facilitated, and where a change in agenda might solve the problem either at the present retrospective or at the next retrospective. These are the structural antipatterns:

Wheel of Fortune …in which the team jumps to conclusions in the retrospective by solving symptoms instead of problems, and the facilitator makes the team members spend time on finding the causes behind the symptoms

Prime Directive Ignorance …in which the team members ignore the Prime Directive—"Regardless of what we discover, we understand and truly believe that everyone did the best job they could, given what they knew at the time, their skills and abilities, the resources available, and the situation at hand" (Kerth 2001)—because they find it ridiculous, and the facilitator reminds them how important this mindset is for a successful retrospective

In the Soup …in which team members discuss things that are outside their power to change, and the facilitator helps them focus their energy on what they can change and accept what they cannot change

Overtime …in which the team gets sidetracked at the retrospective by talking about one development that is not the most important for the team as a whole, and the facilitator helps the team get back on track

Small Talk …in which the team members spend time on small talk in small groups instead of focusing on sharing and learning, and the facilitator changes the activities to make them work together as a team again

Unfruitful Democracy …in which, to the frustration of the minority in the team, democracy is used to decide what to discuss and what to do, and the facilitator finds other ways of deciding that makes everyone happier

Nothing to Talk About …in which the team believes it has become so good that it doesn't need retrospectives, and the facilitator shows the team how it can learn to keep improving

Political Vote …in which the team members wait until the last moment to vote in order to game the system, and the facilitator finds a way to make the voting system more fair

Planning Antipatterns

Planning antipatterns describe problems with the planning of retrospectives. Whom do you invite to a retrospective? Who should facilitate the retrospective? When should you have a retrospective? How much time should you set aside for it? When you find yourself in a planning antipattern, you cannot change the current retrospective, so you need to be aware to plan differently next time.

Team, Really? …in which the borders of the team are blurred, and the team members all help each other figure out who should attend the retrospective

Do It Yourself …in which the facilitator is wearing several hats, which is suboptimal for both the facilitator and the retrospective, and the team finds other facilitators to take over at times

Death by Postponement …in which the team is so busy with "real work" that the retrospectives are postponed again and again, and the facilitator helps the team see how valuable these retrospectives are and that they are real work

Get It Over With …in which the facilitator rushes through the retrospective in order to "waste" as little time as possible for the team, and the facilitator finally decides that to have a decent retrospective, sufficient time must be allowed for discussions

Disregard for Preparation …in which the facilitator initially misjudges how much preparation an online retrospective requires and later learns how to prepare for it wisely

Suffocating …in which team members get tired and hungry and unfocused during the retrospective, and the facilitator makes sure to feed them and give them oxygen so that they can concentrate a bit more

Curious Manager …in which a manager is curious about what happens at the retrospectives and wants to listen in on them, and the facilitator, in a nice but firm way, says no to the manager

Peek-A-Boo …in which team members will not show their faces on the video in an online retrospective, and the facilitator learns why and finds ways to make it safer for people to show their faces

People Antipatterns

People antipatterns describe problems with the people in the retrospectives. You often cannot anticipate these antipatterns because they can occur quite suddenly. Knowing the people will help you be aware of these antipatterns, and the refactored solutions described for these situations can help you navigate out of or around these antipatterns.

Disillusioned Facilitator …in which the team mocks the facilitator for using ridiculous activities, and the facilitator explains why the activities are useful

Loudmouth …in which a team member needs to hear him- or herself all the time, at everyone else's expense, and the facilitator applies various tactics to ensure the rest of the team is heard

Silent One …in which a team member chooses to be almost completely quiet, and the facilitator applies various tactics to make sure the Silent One is heard

Negative One …in which one team member's attitude has great negative impact on a retrospective, and the facilitator shields the other team members from the negativity

Negative Team …in which the team wants to talk only about the negative things because they think these are the only things they can learn from, and the facilitator shows them that a focus on positive aspects can be equally valuable

Lack of Trust …in which the team members do not trust each other enough to share anything of importance in the retrospective, and the facilitator helps them build that trust

Different Cultures …in which the assumptions the facilitator or the team members bring from their own culture are preventing them from seeing how the retrospective is experienced by others, and the facilitator finds ways to make them more aligned

Dead Silence …in which the team members are completely silent, often in an online retrospective, and the facilitator uses various tactics to hear their opinions despite their reluctance to participate

---

Register your copy of *Retrospectives Antipatterns* on the InformIT site for convenient access to updates and/or corrections as they become available. To start the registration process, go to informit.com/register and log in or create an account. Enter the product ISBN (9780136823360) and click Submit. Look on the Registered Products tab for an Access Bonus Content link next to this product, and follow that link to access any available bonus materials. If you would like to be notified of exclusive offers on new editions and updates, please check the box to receive email from us.

Acknowledgments

Foremost, I would like to thank my husband, Erik Corry, for believing in me and letting me spend months without income just to focus on the writing. On top of this, he also patiently (mostly) debugged my entire manuscript more than once. Any errors that remain are, of course, caused by stray cosmic rays hitting my computer. I am grateful to my three wonderful and interesting children, Maja, Sophus, and Vera, for doing their chores so that I had more time to write.

I also want to thank Mai Skou Wihlborg for putting me in a "writer's prison" in her summer house and feeding me so that I could concentrate on writing. She is also the photographer behind the picture on the back cover.

Thank you to Kari Rye Schougaard, Simon Hem Pedersen, Henrik Madsen, and Jakob Roesgaard Færch, who, in the very early days of the book, listened to my ideas and suggested improvements.

I also want to thank the cozy coffee shop Hos Sofies Forældre in Aarhus, Denmark, where I have written most of my chapters while sipping decaf cappuccino or getting a guilty pleasure from eating their wonderful burger with fries and chili mayo.

At Pearson, my editors Greg Doench and Menka Mehta helped navigate many issues in getting this book to publication. And thank you to Martin Fowler for making such a great introduction of me to Pearson.

I want to thank the people who took time to review my book and provide me with feedback, which improved the book immensely: Gary Harvey, Gregor Hohpe, Henrik Bærbak Christensen, Jimmy Nilsson, Joseph Pelrine, Jutta Eckstein, Karl Krukow, Linda Rising, and Toby Corballis. At RFG 2019 Therese Hansen encouraged me to share my unfinished book with the other participants, and their positive reaction made me share the first edition on Leanpub.

This book is masterly illustrated by Nikola Korać, who listened to all my ideas and made something much better than what I imagined in my head.

Finally, a big thank-you to all the people who dislike retrospectives, try to avoid them, and even sometimes destroy them. Without you, this book would not have existed.

About the Author

Aino Vonge Corry is an independent consultant who sometimes works as an agile coach. After gaining her PhD in computer science (CS) in 2001, she spent the next 10 years failing to choose between being a researcher/teacher in academia and being a teacher/facilitator in industry. She eventually squared the circle by starting her own company, Metadeveloper, which develops developers by teaching CS, teaching how to teach CS, inviting speakers to IT conferences, and facilitating software development in various ways. In her youth, she thought she would be a schoolteacher in mathematics, but the strange ways of the world moved her toward communication and facilitation, mainly in the software industry. She has facilitated for the past 15 years, during which time she has made all the mistakes possible in that field. And she would like to share the stories of those mistakes, and their resolutions, as a warning and a guide to other people wanting to facilitate meetings.

Aino has spent time in Stockholm and Cambridge, but she is now back in Aarhus, Denmark, where she lives with her family and a growing collection of plush cephalopods.

To get in touch, email aino@metadeveloper.com, tweet @apaipi, or visit metadeveloper.com.

The Story Begins

At Titanic Softwære A/S, a fictional Danish company specializing in navigation software for ships, changes must be made. The customers have been complaining about bugs in the software that make their ships sail to the wrong destinations, and the speed of software development is not satisfactory.

The chief technology officer (CTO) and some developers decide to go to a software conference to learn what other companies are doing to stay ahead of the game. They find a new approach called agile development, which promises to speed up development, reduce defects, and better align their software with the needs of their customers. In other words, this new approach might take them to where they want to be: rich and happy.

The CTO sends everybody to a scrum master course. Let us zoom in on one little team consisting of six people: Bo, Peter, Rene, Kim, Sarah, and Andrea. Sarah, who was previously the project leader, is the one most eager to become a scrum master, and since eagerness is an important quality in filling this role, she is named scrum master for the team.

Job titles have to be adjusted to the agile way of doing things. Peter, who was the business analyst, becomes the product owner. In her new role as scrum master, Sarah is, among other things, responsible for making sure the daily standup meeting takes place. When issues come up during the daily standup that require talking to others in the company, Sarah makes sure to do that. Also, as scrum master, she is the master of ceremonies and, as such, makes sure that the team has regular retrospectives.

Structural Antipatterns

Wheel of Fortune …in which the team jumps to conclusions in the retrospective by solving symptoms instead of problems, and the facilitator makes the team members spend time on finding the causes behind the symptoms

Prime Directive Ignorance …in which the team members ignore the Prime Directive—"Regardless of what we discover, we understand and truly believe that everyone did the best job they could, given what they knew at the time, their skills and abilities, the resources available, and the situation at hand" (Kerth 2001)—because they find it ridiculous, and the facilitator reminds them how important this mindset is for a successful retrospective

In the Soup …in which team members discuss things that are outside their power to change, and the facilitator helps them focus their energy on what they can change and accept what they cannot change

Overtime …in which the team gets sidetracked at the retrospective by talking about one development that is not the most important for the team as a whole, and the facilitator helps the team get back on track

Part I

Small Talk …in which the team members spend time on small talk in small groups instead of focusing on sharing and learning, and the facilitator changes the activities to make them work together as a team again

Unfruitful Democracy …in which, to the frustration of the minority in the team, democracy is used to decide what to discuss and what to do, and the facilitator finds other ways of deciding that makes everyone happier

Nothing to Talk About …in which the team believes it has become so good that it doesn't need retrospectives, and the facilitator shows the team how it can learn to keep improving

Political Vote …in which the team members wait until the last moment to vote in order to game the system, and the facilitator finds a way to make the voting system more fair

Wheel of Fortune

. . . in which the team jumps to conclusions in the retrospective by solving symptoms instead of problems, and the facilitator makes the team members spend time on finding the causes behind the symptoms

Chapter 1

CONTEXT

When planning her first retrospective, Sarah finds herself short of time. She needs to find a recipe for facilitating one, and her only source of knowledge about how to facilitate a retrospective comes from the scrum master course. The course covered all of the components of the Scrum framework, but its broad scope limited the amount of time spent teaching any one topic, including how to facilitate retrospectives. Sarah decides to do the Start-Stop-Continue activity, where the team brainstorms on what to start doing, what to stop doing, and what to continue doing. She uses this activity as the heart of the retrospective, and in the Decide What to Do phase, the team votes on the three different topics and finds things to start doing, stop doing, and remember to continue doing.

Start-Stop-Continue

For this activity, the facilitator prepares by creating three posters on flip paper or whiteboards, one labeled Start, one labeled Stop, and one labeled Continue. The team members are given Post-it Notes and pens. Some people prefer that all the notes be the same color; others prefer notes of three different colors, such as traffic-light colors, so the notes can be removed from the boards without losing the reference to one of the three choices. The facilitator can also ask people to write Start, Stop, or Continue at the top of the note. The team members are given about 10 minutes to write on the Post-it Notes what they think should be started, stopped, and continued. They then place the Post-it Notes on the boards. The team must now choose their action points based on these, and how that is done depends very much on the facilitator, the content of the Post-it Notes, and the people on the team.

At Sarah's first retrospective, the Start board has a Post-it Note saying "more pair programming." This is an easy thing to address: the team simply makes a schedule requiring everyone to do 3 hours of pair programming each day. Everybody is happy after the retrospective, but in the next sprint, they find that the amount of pair programming being done has not increased significantly. In the first week, Bo and Peter did some pair programming—though not the 15 hours the new schedule called for. Kim and Rene did even less pair programming. Both seem uncomfortable with it, so neither wants to initiate it.

GENERAL CONTEXT

As a retrospective facilitator, you sometimes encounter people with the attitude that a retrospective takes time away from real work. This viewpoint might persuade you to spend as little time as possible on the retrospective in order to make everyone happy, so a quick, fix-it-all activity seems like the best idea.

ANTIPATTERN SOLUTION

The easiest way to do a retrospective is to put up three posters labeled Stop, Start, and Continue. The next step is to have the team brainstorm, write on Post-it Notes, and put the notes on the appropriate poster. The last step is even easier. You start doing the items listed on the Start poster, stop doing those on the Stop poster, and continue those on the Continue poster. The retrospective is finished in 15 minutes, and the team can go back to their "real" work. Even if you added a Do More Of poster, everyone would be back at their desks in short order.

Actually, this approach is not the easiest way to run an effective retrospective. I have heard of "retrospectives" (and I use the term *retrospective* loosely here) where the team went into a meeting room, the scrum master or the project leader asked if anything needed to be changed, and when no one said anything, they all went back to work.

CONSEQUENCES

The benefit of this antipattern solution is that it is a very fast way of conducting a retrospective, but the negative consequences can be severe. Sometimes, you are lucky and the Post-it Notes identify the real problems. To fix them, the team simply stops doing the Stop items and does more of the

Do More Of items. But often, the Post-it Notes merely describe the symptoms of bigger problems that need a more fundamental change than just addressing the symptoms can provide, so you need to stop the symptoms from occurring in the first place by removing the underlying problem.

In our example from Titanic Softwære A/S, this antipattern works fine if the only reason for not doing more pair programming is that people simply forget and revert to earlier habits. However, the team's difficulty may well be a symptom of a problem that is more interesting and harder to solve.

Had they spent time during the retrospective exploring the cause of the problem, they might have found reasons that the team was not doing more pair programming. For example, perhaps the developers couldn't see the benefit of it and had never been asked if they wanted to do pair programming. Or perhaps they saw the benefit but most of them were introverted and thus needed hours on their own to reflect on information before sharing it with their coworkers. Another reason might be that the team members don't know how to do pair programming, so they need to learn how to do it right before they can actually start doing it. On our little team, where Rene is a somewhat negative **Loudmouth** and Kim is a **Silent One**, there could also be a lack of psychological safety. None of these problems can be solved just by forcing people to work together.

SYMPTOMS

If you start hearing things like "Why do we always talk about this at the retrospective?" and "The retrospectives aren't working—nothing changes except the color of the problems," then you should wonder if you are in a **Wheel of Fortune**.

As in a real-world wheel of fortune, you sometimes hit a win when you turn the wheel—if the wheel lands on the actual problem and not symptoms of the problem, that is. And the chance of that happening is about the same as if you spun a wheel of fortune at a fair. Consequently, another symptom of this antipattern is that you hear the same issues being discussed repeatedly at the retrospectives for the simple reason that you never solve the problems, you just put Band-Aids on the symptoms.

REFACTORED SOLUTION

It is tempting to go directly from problems to solutions, and most developers are trained to do exactly that. Unfortunately, since the issues that come up at the retrospectives are sometimes not understood immediately, you need to examine them before you start working on the solutions. After data is collected, you need to look at the causes behind the problems. This is called the Generate Insights phase, as described in the introduction, and it can't be skipped.

Several activities can be applied in this stage. A simple one is to ask for the story behind the Post-it Note in order to learn about what led to this issue. Other activities are the Fishbone (see Figure 1.1 and description) and 5 Hows. Previously, I used an activity called 5 Whys to make the cause analysis, but because of insights I gained from John Allspaw (2014), I now use 5 Hows.

5 Hows

5 Hows is an iterative interrogative technique used to explore the cause-and-effect relationships underlying a particular problem. The primary goal of the technique is to determine the root cause of a defect or problem by repeatedly asking, How did this happen? Each answer forms the basis of the next question. The five in the name derives from an anecdotal observation on the number of iterations needed to resolve the problem.

Few problems have a single root cause. To uncover multiple root causes, the method can be repeated, asking a different sequence of questions each time.

The method provides no hard-and-fast rules about what lines of questions to explore or how long to continue the search for additional root causes. Thus, even when the method is closely followed, the outcome still depends on the knowledge and persistence of the people involved.

Allspaw (2014) explains that asking why can lead to trying to find one cause for a problem and even to blaming. Asking how instead could lead to a narrative, a number of causes. Making it a blame game is obviously not desired, and searching for one cause in a system as complex as software development and teamwork is optimistic bordering gullible. You could even do this on a personal level: try asking yourself how instead of why next time something doesn't go the way you intended.

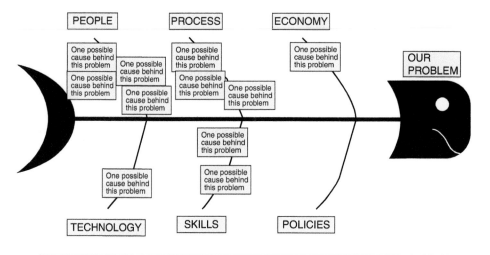

Figure 1.1 Fishbone activity for finding several causes

Fishbone

A Fishbone is a cause-analysis activity in which the facilitator draws a fish skeleton with different themes, such as People, Process, Economy, and Technology, as the ribs to visualize the different types of causes behind a problem. Then the facilitator asks team members to write Post-it Notes for the different causes they can think of and to place them on the fishbone. The result is a fish displaying many different ideas about what might be causing the problem. Perhaps you'll find that most of the ideas cluster in a particular category, such as process or technology, and that gives you input for what cause to work on. This technique is generally used in the Generate Insights phase of a retrospective. It is also called an *Ishikawa diagram analysis*.[1]

1. https://en.wikipedia.org/wiki/Ishikawa_diagram

All Generate Insights activities for finding causes involve digging down into what is written on the Post-it Notes to find the real stories behind them.

Every meeting, including a retrospective, has a life cycle of divergence and convergence (see Figure 1.2), as described in *Facilitator's Guide to Participatory Decision-Making* by Sam Kaner (2007). We start, hopefully, in a small space of understanding the purpose of the meeting and the outcomes we expect; at the least, the topic of the meeting should be agreed upon. Then we go into the divergent phase, where discussions open up for disagreements and clashes of opinion. Next comes the groan zone, where people further discuss the issues not in search of agreement but to learn more about the topic. After spending some time in the groan zone, the meeting needs to go into a convergence phase where we try to narrow down the scope and/or find agreement. If agreement cannot be reached, at least the group can agree to disagree and establish the terms and grounds of the disagreement.

To start with, the agreement on the topic resembles the Set the Stage phase of the retrospective. The divergence part of a retrospective is the Gather Data and Generate Insights phases. It is not an easy part of a retrospective but a very important one. The groan zone can be related to the last part of the Generate Insights phase and the start of the Decide What to Do phase. Convergence is the essence of the Decide What to Do phase, and when you reach the decision at the end of the retrospective, you Close the Retrospective. Make sure to go through all five phases of a retrospective and generate insights instead of jumping to conclusions and thus getting premature convergence just to get to the point quickly.

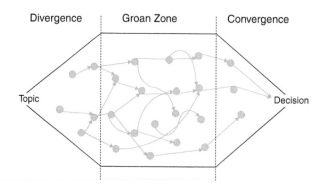

Figure 1.2 Divergence and convergence in a meeting

ONLINE ASPECT

If the retrospective is done online, you can use an online tool that forces the team to enter and stay in the Generate Insights phase before they move on to problem solving. If you are creating your own online document, you can visualize what phase of the retrospective you are in at all times. I often use a Google Drawings diagram for the Generate Insights phase and cross out the suggestions box. In that way, team members are visually reminded not to add a suggestion for an experiment before we get to the Decide What to Do phase (see Figure 1.3).

You can also visualize the phase of the retrospective by wearing a different hat for each phase. If people still forget not to jump to conclusions, you can point at your hat and smile.

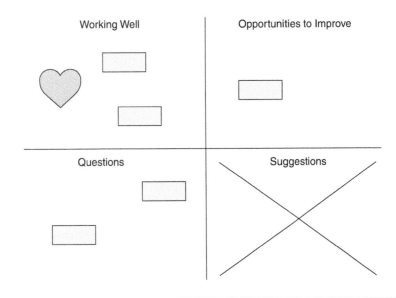

Figure 1.3 An online retrospective document with the suggestion box crossed out

PERSONAL ANECDOTE

I was facilitating a team in a company in Denmark that had been acquired by a large organization from abroad. The company had its own way of doing things and liked it like that. Unfortunately, the new organization decided to divide the original company into three teams. At one of the retrospectives not long after the acquisition, one of the things that was put on the board was "less work in silos." We all understood what that meant: the team members wanted to work together instead of as three separate teams working in parallel without helping each other.

Since we understood the issue, it was the action point they decided, without much discussion, to do at that retrospective. And at the next retrospective. And the next retrospective. Finally, we decided to look at the cause of the siloing. It turned out the key performance indicators made for each team got in the way of cooperation. Since every team had its own goals, it became a zero sum game where one team's win was based on another team's loss. Also the planning meetings for the teams were with different people in the new organization. It would have been good to acknowledge that from the beginning, to understand whether this was something they could change or something they would have to learn to live with. (See also Chapter 3, **In the Soup**.)

Now for something completely different: working mainly with IT people, I always find it hard to make people stay in the open discussion, or the groan zone. It feels almost physically hard to keep the options open before I allow them to rush to the solution phase. All minutes spent in the open discussion phase are well spent, though, in my experience. This is where new insights and ideas often come up. Dave Snowden has written often about premature convergence as a part of complexity theory, and he describes what we might miss out on: "holding things open, allowing them to break down into more finely grained objects, then seeing them recombine and co-evolve" (Snowden 2015).

When I had considerable experience with IT teams, I was asked to do idea generation for people working at an art museum. I was not prepared for what happened; it was extremely easy to get them into the open phase where everything is possible and where they should not fixate on a solution or track. I felt that it was going very well and that I had done a good job of getting them into that way of working.

But then my problems started. It was next to impossible to get them to make decisions and to cut down on ideas and remove possibilities. I found myself running from group to group and peeling down Post-it Notes, which they put up again the moment I turned toward another group. When I heard someone say, "Should we even have an art museum?" I felt a slight panic, since we were here to brainstorm for an app, not have an existential discussion of whether we needed an art museum. I had to adapt to this new crowd, and I had to do it quickly.

I interpreted the situation as working with a different species, almost. I had to be much more direct than was necessary with other groups I'd worked with, and I had to spend more energy in making this group go to the last phase of a meeting and get to some kind of convergence. In the end, we succeeded in agreeing on a focus for the next period only by promising that the other ideas would be kept for the future. It is interesting how cultural this behavior, or thought pattern, is. This is also described in Chapter 23, **Different Cultures**. In some cultures, dealing with conflict or differences of opinion is something you try to avoid. People from such a culture (and it could be within a specific country, company, or team) will seek consensus as quickly as possible, thus perhaps losing the chance for novel ideas to arise from their interaction.

Prime Directive Ignorance

...in which the team members ignore the Prime Directive—"Regardless of what we discover, we understand and truly believe that everyone did the best job they could, given what they knew at the time, their skills and abilities, the resources available, and the situation at hand" (Kerth 2001)—because they find it ridiculous, and the facilitator reminds them how important this mindset is for a successful retrospective

Chapter 2

CONTEXT

For her next retrospective, Sarah has done her homework. She has read not only *Agile Retrospectives: Making Good Teams Great* (Larsen & Derby 2006) but also *Project Retrospectives: A Handbook for Team Reviews* (Kerth 2001). She discovers that there is a lot to think about when planning and facilitating a retrospective, but luckily, Larsen and Derby's book has made planning an easier task by providing suggestions for entire retrospective agendas.

With a planned retrospective ahead of her, Sarah is excited and a little bit worried. The thing she most worries about is how to present Norm Kerth's Prime Directive (2001) to the developers, since she believes they might find it ridiculous. In the end, she decides to drop it.

Prime Directive

Regardless of what we discover, we understand and truly believe that everyone did the best job they could, given what they knew at the time, their skills and abilities, the resources available, and the situation at hand. —Norm Kerth

The last sprint had been a disaster with everyone working overtime and no one feeling good about the result. Well, almost everyone worked overtime—Peter had not. He went home every day at the usual time; he would sneak out thinking nobody noticed him and arrive late in the mornings. There had been some grousing about this by the water cooler and in the corners, and it was secretly decided that the retrospective was the place to deal with Peter's seeming lack of commitment to the team. Sarah and Peter were the only people who did not know of this plan.

At the retrospective, they looked at the data—the regression tests and the feedback from the demo with the users—and they put Post-it Notes on a timeline to share their experiences. It was obvious that Peter was being blamed for the team's overall dissatisfaction. There were Post-it Notes with his name on the board, and he became even more silent than he usually was. Since there were so many notes, Sarah decided to read them all out herself, thinking it would take too long if they all had to read a portion. But there was also another reason Sarah chose to read the notes: as she read them aloud, she tried to avoid the Post-it Notes with Peter's name on them. Everybody had already seen the notes, though, so Sarah's effort to shield Peter was in vain. Peter left the retrospective in complete silence, and the rest of the team members felt unsure about what to do next. They had hoped that he would apologize to them or explain his behavior, but they had never really given him a chance to enter the discussions.

Timeline

In this activity, the facilitator prepares by drawing a timeline on the wall of the retrospective room. This can be done either by writing dates or events on Post-it Notes and placing them in a line or by actually drawing a line on a big whiteboard and adding the start and end dates. The team members are asked to think back on this time, from beginning to end. They will be given Post-it Notes in different colors, red for events that took away their energy (made them sad/unhappy/mad), green for events that gave them energy (made them happy/relieved), and yellow for events that puzzled them, for questions, or for events that are both good and bad.

Then the team and the facilitator look at the board and the colors and decide which parts to start discussing. For an explanation of why this discussion is important, see Chapter 1, **Wheel of Fortune**. The purposes of using a timeline are that some people find it easier to have that framework and that a sprint or an entire project development phase can be seen as an overview with the colors indicating the happiness/energy at different times.

Sometimes, the occurrence of this antipattern is a symptom of another antipattern: **Lack of Trust** (Chapter 22). In these cases, the problem is deeper than ignorance of the Prime Directive and cannot be fixed simply by emphasizing Kerth's words of wisdom. The real problem in the anti-pattern solution for **Prime Directive Ignorance** is the lack of trust that makes it impossible for Peter to share what is going on with him and for everybody else to expect that he is doing his best under the circumstances he is in.

GENERAL CONTEXT

Kerth was the first to write about retrospectives, and he coined the term after they had been called postmortems for some time. He wrote the Prime Directive, and it has been the center of much discussion in the IT community:

Can we really, regardless of what we discover, *understand* and *truly believe* that everyone did the best job they could? Even if we made allowances for what they knew at the time, their skills and abilities, the resources available, and the situation at hand?

At the end of a project, everyone knows so much more than at the beginning. Naturally, we look back on decisions and actions we wish we could do over. This is wisdom to be celebrated, not judged and used to embarrass people. Adhering to the Prime Directive, we want to combat confirmation bias in retrospectives. If we are filtering information to accept only data that supports our preconceived opinions, we are missing a chance to learn.

The problem is that it feels awkward to follow this directive, because it is hard to *really* and *truly* believe that everybody else is doing their best when you *know* they are slacking or being lazy. One of the reasons that we find this hard is explained by the *fundamental attribution error* described in *The Intuitive Psychologist and His Shortcomings: Distortions in the Attribution Process* (Ross 1977). This error explains how we attribute flaws in the behavior of other people to internal traits, such as laziness, stupidity, and so on, instead of attributing them to the circumstances this person is in. As an example, if you are late for a meeting, you might attribute your tardiness to the problems you had with getting the kids to school or to the bus being late. But if someone else is late to the meeting, you might roll your eyes and think he or she is an irresponsible person, according to the fundamental attribution error.

In the IT industry, which is the one I have worked in the most, there is a focus on errors (bugs) and bad work (which carries legacy bugs and poor programming to future iterations). Maybe this focus on the negative exists because of the need for precision when working with computers, and perhaps this need attracts perfectionists to the IT trade. Perfectionism can be a good thing for most of what we do, but it is not helpful when we are trying to understand why things happened and find sound solutions to problems.

ANTIPATTERN SOLUTION

Just forget the Prime Directive. It's too touchy-feely for a logical bunch of programmers. As simple as that. Ignore it. Or recite it with a mocking smile on your face, signaling that everybody can just forget about it. Go ahead with the retrospective and get on with the data collection.

CONSEQUENCES

Your retrospective will now be like any other review session, where we, like most people, are interested only in finding a scapegoat to direct blame away from ourselves. Scapegoating is a common nonsolution to problems, and you probably experienced it in childhood when your parents asked you and your siblings, "Who started it?" or "Who broke the vase?"

The consequence could be that participants bring all their assumptions and negative expectations to the retrospective instead of anticipating a chance to share and learn. They do not really listen to what others are saying because they already believe they know whom to blame. The retrospective can easily become a blaming and shaming session, where the people being blamed are afraid to share and in the end refuse to attend the retrospectives because they become too painful.

SYMPTOMS

Team members are afraid to go to the retrospectives when something has gone wrong. They are not eager to learn to become wiser but instead are afraid of what might happen at the retrospective. People arrive in a defensive mood, armed with angry counterarguments to the attacks they expect and unwilling to explore the issues together.

REFACTORED SOLUTION

Bring the directive to each retrospective, and start with it. I have found that some groups react negatively to the wording of the Prime Directive, and in those cases, I reword it but maintain the core idea that we should all look for the problems in the system, not in the people. One way to put the Prime Directive front and center in everyone's minds is to send an email before the meeting explaining that retrospectives are based on the assumption that everyone did the best they could under the circumstances and that team members are obligated to respect this assumption. Try this approach, and you will likely lay a foundation for building trust among team members.

The Prime Directive should be seen as a way to think about the assumptions and expectations we have about other people, a state of mind to enter when you walk into a retrospective. It is a team thing to have a retrospective, so you should think of everyone as part of the team, or as part of a system, and find ways to improve that system. If you start out by thinking they purposely have not done their best, you will not have a fruitful retrospective.

Remember what Yoda said to Luke when Luke said he did not believe he could lift his spaceship out of the swamp: "That is why you fail." Yoda succeeded, after some time, in putting Luke into the right mindset when he tried to tackle problems.

Had our Danish team entered the retrospective thinking about the Prime Directive, they might have learned that Peter's wife was terminally ill and that this was the reason he was not pulling his weight at work. Maybe Peter would have shared, had he felt secure and safe at the retrospective. Then they might have found something that Peter could do together with someone else or given him less work for a period while he figured out what would happen to him emotionally and financially as he went through this very difficult and painful part of his life.

It could also be that Peter was uncomfortable sharing this at a retrospective and needed a more private setting to share this personal information and its consequences. In any case, accepting that Peter was not to blame and acknowledging that the way the team worked together and communicated was faulty would have been a more likely and positive result had they focused on the mindset proposed by the Prime Directive.

ONLINE ASPECT

If the retrospective is conducted online, you can add the Prime Directive to the email that is attached to the calendar invitation. You can also have the Prime Directive on a poster behind your head as a constant reminder of it. You might also be able to send messages directly to the people you think are not following the directive to remind them about it without making them lose face in front of everyone else.

PERSONAL ANECDOTE

Many years ago, in a retrospective not facilitated by me but for my team, something very sad happened. During the data gathering, one name was singled out on the board with some negative comments. The facilitator did not stop it, and soon a whole swarm of Post-it Notes about that person filled the board. Not surprisingly, this experience was unpleasant for the person in question, but it was about a behavior that everyone was tired of and that affected our work together. The facilitator allowed us to start talking about the different Post-it Notes on the board, what they meant, and the stories behind them.

The person in question started defending himself, but in the end, he stormed out of the room. We never got anywhere near understanding what actually created the behavior or what we could do collectively to help this person or at least learn to work around it. The time spent on that retrospective was more than wasted—it harmed us for good.

Another time, I witnessed something touching concerning a young man who had recently joined the team. I could see from the start of the retrospective that he felt really bad about something. It turned out that he thought he had done a disastrous job of working on the front end of the system, taking too long, making mistakes.

But the rest of the team turned to him, when they realized how he felt, and gave him a veritable love-storm. They let him know that of course he needed time to learn about the system. In addition, they told him that because of his work, they had learned which parts of the system needed more documentation or even a rewrite. The young man was so happy and relieved, it almost brought a tear to my eye!

In the Soup

... in which team members discuss things that are outside their power to change, and the facilitator helps them focus their energy on what they can change and accept what they cannot change

Chapter **3**

CONTEXT

The team is now really comfortable with retrospectives, and Sarah is happy about this progress. They have had many discussions and conducted many experiments, and the team can see cooperation, code quality, and overall happiness increasing over time. At the last two retrospectives, alas, the team ended up choosing to discuss the missing test framework. When they dot-vote, this issue is always voted as the most important problem to solve.

Unfortunately, these discussions always lead to the same conclusion: that management resembles a stingy Uncle Scrooge. The team ends up discussing prices, possibilities, technologies, and management, but they make no progress toward a solution. Sarah knows the missing test framework will most likely be the subject of the next retrospective as well. She is as frustrated as the rest of the team about this issue.

Dot Voting

Dot voting, also known as *dotmocracy*, is a method used to vote on different topics in order to quickly decide what to discuss or what to do. Topics are written on Post-it Notes, and team members vote with dot stickers or marker pens. People have a limited number of votes to cast, and after the votes are in, the "winner" is the one or more topics with the most votes.

Some problems can occur with this method; they are described in Chapter 6, **Unfruitful Democracy**, and Chapter 8, **Political Vote**. Some facilitators have stopped using dot voting, due to the reasons described in these antipatterns.

GENERAL CONTEXT

It is often the case that, after clearing the small issues during the first round of retrospectives, the team encounters a real hurdle—a problem that is out of their domain of power. They might not realize at first that this problem is out of their hands, and they happily spend hours discussing solutions they have no power to implement. After a few retrospectives, however, the discussions become more frustrating than happy.

ANTIPATTERN SOLUTION

A subject that needs management approval/action can become the starting point of many frustrating retrospectives, since the team cannot resolve the problem. The retrospectives turn into groundhog days,[1] where the same issue seems to be discussed from different angles at every meeting.

The antipattern solution is simply to do what you were taught initially as a retrospective facilitator: you let the team decide what the biggest obstacle is and let them discuss it in search of a solution. It seems like the right decision, because retrospectives are about reflection and the team finding the best solution themselves, but in some cases, it can lead to the negative consequences described next.

CONSEQUENCES

When the team encounters a situation they cannot change, it is sometimes because they lack the skills to do so, but more often, it is because they lack the authority to do so. Retrospectives spent looking for solutions that the team has no power to act on are doomed to unfruitful outcomes. In addition, team members may form a negative view of retrospectives in general, seeing them as an unproductive use of their time.

Sometimes, of course, what the team wants aligns beautifully with what management wants, but at other times, management has different priorities, and the situation stays the same. Retrospectives degenerate into complaint sessions and are a waste of time.

---

1. Referring to the film of the same name, where a man is destined to relive the same day over and over again, ad nauseam, until he learns to be a better person.

SYMPTOMS

You hear people make comments such as "We want to work on the things that really matter, not just which coffee we should buy" or "We are always discussing the same thing."

REFACTORED SOLUTION

After the "problems" are collected, draw two circles on the board, as shown in Figure 3.1. The innermost circle contains issues that the team can control—that they expect to have the power to change (e.g., "start doing code review" or "change location of standup meetings"). The next circle contains issues over which the team has some influence. It is for issues the team cannot directly change but for which they can take persuasive action.

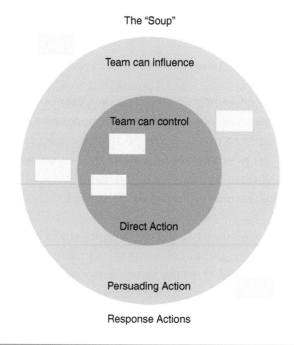

Figure 3.1 The Circles and Soup activity

Typical issues in the outer circle are the way the management asks the team to start new things or how the team cooperates with another team—issues related directly to the team in one way or the other but that involve people not present at the retrospective.

The space outside the circles is endless, so a third circle is not used. This space contains all the issues that are part of the environment, "the way it is," or the *soup*. These are issues that affect the team but that the team has no influence over (e.g., the company is losing money and has to lay off employees). Other issues in this space might be the geographical placement of the offices or the personality of someone in the company.

This activity is called *Circles and Soup* (Larsen & Derby 2006), but it is also known as *circles of influence*. The activity is to be used much like the Serenity Prayer: "God, grant me the serenity to accept the things I cannot change, courage to change the things I can, and wisdom to know the difference."[2]

What makes the Circles and Soup activity productive is that it gives the team a realistic perspective of the issues they want to resolve and the scope of their power to address them. As they examine the causes of "soupy" issues that seem beyond their control, they sometimes find that those issues can be moved into the circle of influence or even into the circle of control. For example, maybe the team wants to sit closer to their testers in India, but a cause-analysis activity such as 5 Hows or Fishbone reveals that the real reason for needing more communication is that the initial communication is insufficient or that the documentation received from the testers is misunderstood. And now it is suddenly an issue the team can do something about.

---

2. The prayer is often attributed to the Protestant theologian Reinhold Niebuhr, although he conceded he might have been influenced by a forgotten source.

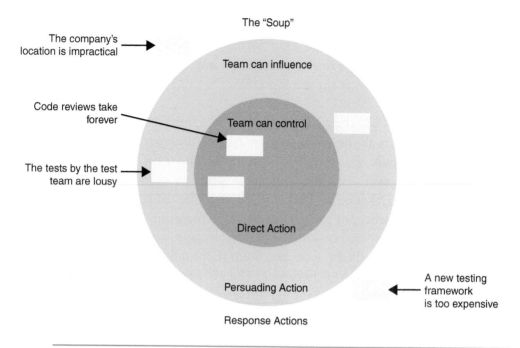

The "Soup"

The company's location is impractical →

Team can influence

Code reviews take forever →

Team can control

The tests by the test team are lousy →

Direct Action

Persuading Action ← A new testing framework is too expensive

Response Actions

Figure 3.2 The Circles and Soup activity in Gather Data phase

For the issues that remain in the soup, it is sometimes beneficial to invite someone from another team, or perhaps a manager, to the retrospective. However, as discussed in Chapter 15, **Curious Manager**, "outsiders" may inhibit the team members' participation. In general, my viewpoint is that the only people you invite to a retrospective are the people in the core team. But if the Circles and Soup activity shows you that some important issues would benefit from having someone outside the team present at the retrospective, I would ask the team if they would object to inviting that person to the next retrospective. This goes for managers as well as other teams' members.

If it is not possible, due to geography, to have the other person present in the room, you will have to arrange an online retrospective. A whole new set of retrospective antipatterns awaits there, though (e.g., see Chapter 16, **Peek-A-Boo**, and Chapter 24, **Dead Silence**).

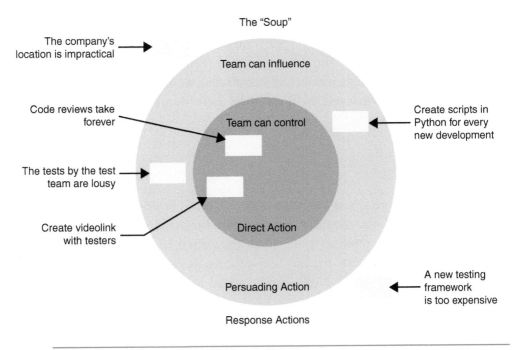

Figure 3.3 The soup in Decide What to Do phase

Another solution to avoid always talking about the same issue is to be strict in the voting process—for instance, by allowing only one vote on one subject from the outer ring, the soup, thus focusing on the issues the team can actually do something about.

As an alternative, you can ask the team how they can make the issue in the soup worse. If they can make it worse, they can also make it better, which shows they have some influence. This approach is called *paradoxical intervention* and is used in therapy and coaching to make people aware of their free will. It is often dismissed as reverse psychology, but where reverse psychology is used to manipulate people into doing what the manipulator wants them to do (by proposing the opposite), paradoxical intervention is meant to support the person's ability to make choices by making them aware of the power in the situation.

I sometimes use an approach like this in my computer science teaching when the students have no idea what to do in an assignment. I ask them what they definitely should not do, and then we can start a discussion from there, because when they know what would not work, they can move toward what will work. It kicks their brains out of the deadlock they are in.

ONLINE ASPECT

If the retrospective is online, it is not practical to do the Circles and Soup activity on a whiteboard. I often prepare the activity in a diagram and send everyone a link to it. Then, all the virtual Post-it Notes can be copied to that diagram and divided into sections, or else one note representing each group of Post-it Notes can be added to the Circles and Soup diagram.

PERSONAL ANECDOTE

In one team, the testing was an issue. The team wanted automated tests of the entire system, and they wanted to buy testing software cooler than Cucumber or perhaps assign a person exclusively to operating Jenkins. Management did not see the point of spending money on new software, and at every retrospective, at least 5 minutes were spent discussing the issue. Now, it is not an uninteresting issue to discuss, but when nothing can be done about it, time is better spent figuring out how to learn to live with this fact.

Once we tried the Circles and Soup exercise, the team understood that this was indeed one of the issues that are better treated as a fact of life and that time was better spent discussing alternative solutions. The team invited the manager to the retrospective to try to make management understand how big an issue testing was and what impact it had on agility, speed of development, and time to market. Unfortunately, the manager did not understand the problem, and the team decided to learn to live with it.

In the end, the team members decided they could build the tests themselves in Python, little by little, and get the tests exactly right, so they would no longer be afraid to change the code. They could do it without management's approval to spend more money on the testing framework. Obviously, their extra work also had a cost, but this one could be hidden from management.

Of course, the whole issue was a symptom of a larger problem: the team's relationship with management was poor. This relationship could have been improved by better communication, understanding, and respect between the team and management. In an ideal world, the team would not feel the need to hide things from management in order to do their best work, but sometimes the world is not ideal, and we have to make do with what we have.

Overtime

. . . in which the team gets sidetracked at the retrospective by talking about one development that is not the most important for the team as a whole, and the facilitator helps the team get back on track

Chapter 4

CONTEXT

In the next retrospective, an unexpected issue came up. Andrea heard that their immediate boss, Nancy, had decided to leave the company. This was a huge blow for the team because even though she had been their boss for less than a year, the team had been very happy with her. In contrast to their previous boss, Anna, Nancy listened to the team and tried whole-heartedly to help with every issue they brought to her attention.

The team members felt that Nancy understood them and cared for them. She often showed her thoughtfulness through gestures such as bringing homemade cake to staff meetings and remembering everyone's birthdays. Therefore, it was not without reason that the team was upset to hear she was leaving. The exception was Rene, who kind of liked the direct, almost harsh style of their previous boss. The team spent a lot of time discussing Nancy's departure, and Sarah felt they really needed to process the situation together as a team.

Since Sarah was part of the team, this might also be one of the cases of the **Do It Yourself** antipattern (see Chapter 10), where the facilitator is also part of the workings of the team. But, after the initial discussions, Sarah decided to move forward with the retrospective. She had a plan and she wanted to follow it.

Sarah noticed early on in the retrospective that the team was not following her planned time schedule and that they would go over the allotted time. She asked if they could stay for 10 minutes more, but she had been too optimistic and it went over by 20 minutes. Before she closed the retrospective, two people had left the room due to other engagements, and two others, thinking the retrospective was drawing to a close, had allowed themselves to be distracted by emails and chats and were focusing on their phones.

GENERAL CONTEXT

An overtime problem can arise at a retrospective for various reasons, such as when a new and interesting incident occurs and has important causes and ramifications. The facilitator and team are naturally eager to discuss the matter, but the facilitator also wants a real retrospective. He or she does not want any team members to feel that the retrospective was unproductive.

Although new incidents or issues are a common reason that retrospectives go over time, other factors can also push the meeting past schedule. For example, the facilitator may find it difficult to stop a particular discussion, or the team may lack energy and have a hard time moving from one activity to another.

ANTIPATTERN SOLUTION

The facilitator agrees the team needs more time in the retrospective and extends it by 10 minutes. The extension, however, turns out to be inadequate, and the retrospective becomes even longer. Often, when a new issue comes up, it is more significant for some team members than for others. For some, it is the most important thing going on in and around the team, but frequently, other team members stop participating because the new subject has little or nothing to do with them.

CONSEQUENCES

Since the discussion most likely involves only two or three people, the rest of the team will feel stuck listening to something that may not concern them at all. They want the problem solved for the other team members, of course, but maybe they had another problem they wanted to address or other tasks they need to get done, so they feel this retrospective is a waste of their time.

Also, if the retrospectives are allowed to go over time, it becomes increasingly challenging to convince people to spend time on them, since they will be afraid that 1 hour actually means 2 hours. In effect, not everybody from the team will be present; they will either be physically or mentally absent.

SYMPTOMS

The retrospective goes over time, and even with added time, it runs past the extension.

REFACTORED SOLUTION

For every retrospective, you, as facilitator, must have an agenda: a plan for how to spend the time. You should also have a backup plan (that you never show the team), either written next to the official one in your notebook or, if you're experienced, kept in your head.

In this alternative plan, for example, you can choose to spend more time on a specific subject or to spend less time on debriefing if you feel this debriefing was already done during an exercise.

Debriefing

You could call the entire retrospective a debriefing, but in this case, it is the practice, after each activity in a retrospective, of asking the team questions such as "What happened here?" or "What can we learn from this?" or "Who should we tell about this, and why?" The purpose of a debriefing is to help the team step back and look at what they just did in order to generate the most learning and understanding from the activity.

So, changing plans and spending more time than planned on a topic is quite all right and is actually a mark of a good facilitator. The trouble comes when you let it get out of hand and the retrospective goes over time.

When you let people talk more about a subject, you have to be able to calculate how much more time you can give them. Can you shave some time from the **Decide What to Do** phase by letting the team members discuss the suggested experiments in smaller groups, so that they can be discussed in parallel instead of sequentially? Remember to sum up in plenum afterwards so that people do not feel they missed something by not being part of that discussion. If you do it to save time, you have to be aware that the summing up can take almost as long as the plenum discussion you tried to avoid if you do not facilitate the summary. Can you drop a planned vote, because this discussion must take priority? Can you hold a vote about whether they want to stay 20 minutes longer? Does the vote have to be anonymous to get honest answers from polite people? Can you stick to the new timetable?

Anonymous Vote

An anonymous vote can be done in many ways. If it is online, everyone could chat the facilitator their vote with a direct message, or, if the online system supports it, that system can be used to collect the votes. In real-life retrospectives, the facilitator can gather votes on Post-it Notes or index cards. It is important, before every vote, that everyone agrees whether it is to be a majority vote or whether one person can veto it. In the latter case, one person should be able to veto that the retrospective goes over time. Otherwise, that person would have to leave, and it can ruin a team retrospective if one person is not able to be there the whole time.

If you cannot get more time for a discussion, you have to cut it short by saying, "This is a very interesting discussion. I think we should make a note for continuing this somewhere else, some other time." Alternatively, you can have a "parking lot"—a poster where you write the topics that the team wants to discuss but that you have no time for in the retrospective. You have to remember to address the issues in the parking lot afterwards; otherwise, people will stop using it.

If overtime occurs often, you could consider planning for longer retrospectives. People are different. I assume you need more time in Italy than in Finland.[1]

I am often asked how long a retrospective should be. It depends on the size of your team, the amount of time you are looking back on, and the ambitions for the result of the retrospective. If you have a team of six to eight people, 1 hour for each week you are retrospecting over will be sufficient. This means 2 hours after each 2-week sprint, for instance. If you have more people, you will need more time, at least if you want to get the same result. Most often, I spend 1.5 hours every 2 weeks on a recurring retrospective with a team that knows me, because they prepare mentally before the retrospective and we agree on what we want to get out of it. In an online retrospective, it is hard to make it work for more than 75 minutes unless you have a break in the middle. Ideally, no meeting should be longer than 45 minutes without a break, according to research by Daniel Kahneman (2013), but it can be almost impossible to convince people to do that.

I once facilitated a retrospective for 80 people in 1 hour, but the ambitions were much lower than what I normally have. We wanted people to share their experiences on a huge timeline (for a week-long conference) and look at each other's answers. I went around and found some common and some different experiences and summarized them at the end of the retrospective. There was reflection for the attendees, there was sharing of how an incident can be experienced differently by different people, but there was no time to generate deeper insights and learn a lot. But then again, that was not the purpose.

---

1. In my experience, people from different cultures have different verbosity. In Denmark, we often say, "Speech is silver, but silence is gold." In fact, that might be all you could hear a Dane say in an afternoon. Finnish people, perhaps because they live farther north, find even Danish brevity to be a bit much. They think "it" should be obvious and you do not have to talk about it. In fact, this well-known joke is just normal behavior in Finland: A Finn and a Swede sit down to drink. Silence. The Swede says "Skål" (cheers). The Finn answers, "Are we here to talk or to drink?"

ONLINE ASPECT

If the retrospective is done online, it is a bit more difficult to change your agenda, since you are bounded by the online tools you are using. In addition, you are less able to use your body language to change the pace of the retrospective. However, you have other tricks you can use. It is much more accepted in an online retrospective to ask everyone the same question, because people are aware that only one person at a time can speak in an online meeting. You can ask everybody, one by one, if a particular topic is what they want to discuss right now. Or you can ask them to send you a message if you think they are not comfortable enough to share their possibly opposing answer with the group.

PERSONAL ANECDOTE

In the beginning of my facilitation journey, I once facilitated a retrospective in the company I worked for. As it goes in this antipattern, one of the issues that came up turned out to be something people—well, at least, half the people—really needed to talk about. A lot. And one of the people eager to discuss the issue was my manager. See also the antipattern **Curious Manager** (Chapter 15).

I was unable to stop the discussion. I tried several polite moves, and in the end, I decided that this was what they needed to discuss now, and I made time for them to do that by changing my plan for the retrospective. I facilitated that discussion as best I could, and after they came to a conclusion, they turned to me and asked, "What happened to the retrospective?" It was as if they somehow woke up and noticed that an hour had passed. I now had 30 minutes to make a retrospective, which was less than I thought I would have when I made the new plan. I tried, and at one point asked for 15 more minutes, which I got.

It was not enough. After the added 15 minutes, we were in the middle of generating insights, and I had to tell them that I would take the issues they had and use them in the next retrospective. This was not my proudest moment, because the insights I had hoped for them to achieve during the retrospective were lost. I had given them a bad retrospective experience, because they did not get out of it what they should have—namely, reflection together and improvements everybody believed in.

Since then, I have gotten better at keeping retrospectives on track. But 9 years later, I encountered a similar situation at a client's retrospective. I decided on the spot to let the team vent for 30 minutes, then stop the discussion and do a *positivity retrospective*, where things are easier to discuss and less time is spent on deciding what to do.

Positivity Retrospective

A positivity retrospective is one in which the team gets a chance to focus on their strengths, successes, and positive events. The whole retrospective is about sharing powerful and positive events and thoughts, and thus it often starts with appreciation expressed among team members and ends with good wishes for the future. The takeaways from a positivity retrospective are about how to increase the good things. A positivity retrospective is part of the refactored solution for the **Negative Team** antipattern (Chapter 21).

From TCI—theme-centered interaction—we know that disturbances and passionate involvements take precedence, which means that when you work with human beings, issues or incidents sometimes arise that cannot and should not be ignored. It depends on the context whether you should postpone the retrospective in those situations. Based on my experience, it is important to choose and not try to do both in the limited time of the retrospective.

Small Talk

... in which the team members spend time on small talk in small groups instead of focusing on sharing and learning, and the facilitator changes the activities to make them work together as a team again

Chapter 5

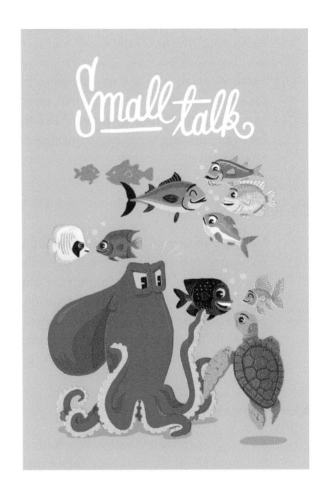

CONTEXT

Titanic Softwære A/S has planned an off-site for next week in Trysil, Norway, where everyone will enjoy skiing together. Most of the people in the company are excited about the trip, and it is on their minds most of the time, especially now that it is only days away. Naturally, when time is set aside for meetings, it is tempting to discuss the trip: Will there be enough snow for skiing? What will they have to pay for themselves—the equipment? the food? the beer? And so on.

Sarah wants to start the retrospective, but she also understands the desire for small talk about the skiing trip, even more so now, because the last sprint was frustrating for most of the team members. She is therefore torn between allowing the skiing trip small talk or ending it and starting the retrospective.

GENERAL CONTEXT

Some people are prone to spend time on small talk. To them, it is a way of relating to others, showing they like other people, and getting attention. Sharing pleasantries is an important way for most people to relax in each other's company, and small talk can start and flow easily between most people because of its innocuous nature. Small talk can be a good thing, especially for teams with remote people, because talking about food, travel, and so on, helps to build trust (see Chapter 22, **Lack of Trust**).

ANTIPATTERN SOLUTION

As a retrospective facilitator, you want to be nice and you want people to feel good about the time they spend with you, so you allow time for small talk. This is not necessarily bad; if time is set aside for it in the retrospective and if everybody feels the need for it, small talk can set a comfortable atmosphere for the retrospective. Ending a nice conversation can be daunting, particularly for an unskilled facilitator, and you have to be sure that what you have planned for the team will be of more value to them than their conversation. Too often, the small talk is allowed to continue.

CONSEQUENCES

Time spent on small talk takes time from focused discussions or reflection on issues and events in the team's life together. Small talk can be important, because it helps people bond and provides a way for them to get to know one another. But in the setting of a retrospective, it eats precious time set aside for other discussions.

SYMPTOMS

The symptoms of this antipattern are obvious, since it is hard to ignore people small-talking in the corners of the retrospective. Another, even more serious, symptom is that the facilitator runs out of time for making it a fruitful retrospective, for example, by skipping a phase or by not allowing everyone to talk.

REFACTORED SOLUTION

If this is the first time you encounter small talk among the group, you can merely say that we need to focus on the issue at hand. Changing the activity is also effective, because small talk sometimes occurs when people have spent enough time on a particular task and are getting antsy. If the small talk occurs during small group sessions, it could also be that some people work faster than others. In that case, you should be able to give the chatty group another issue to consider to keep them engaged with the retrospective.

If small talk is an ongoing problem, you should address it at the beginning of the next retrospective by setting *ground rules*. Be sure to ask the team what additional ground rules they would like to have in place at retrospectives and perhaps also in general. Consider also whether small talk is occurring because the team simply needs a break after 45 minutes.

Ground Rules

When a group of people chooses to become a team, they have to accept that they likely are a heterogenous group with various expectations for working together. Some people need to discuss everything in order to learn new things; others need to read in solitude. Some people find it hard to concentrate without music; others need silence. Some people are relaxed about meeting times; others get very annoyed by latecomers.

A team can choose to set ground rules based on their expectations about group work. Some ground rules might apply to the group in general, such as "We are not late to the meetings" and "We do not interrupt each other." Others might be personal, such as "If John is wearing headphones, he does not want to be disturbed."

With ground rules, issues such as small talk don't have to reach the point that you are irritated—you can just refer to the ground rules that everybody on the team agreed to follow. In my experience, a team with ground rules can avoid a lot of negativity and passive aggressiveness. Also, some of the unwanted behavior can be avoided by the facilitator, but it is more efficient if there are more people keeping an eye on the ground rules.

If people continue to engage in small talk even though you have ground rules, you can change the subject or give them more to do in order to change their focus. If the small talk continues, try moving closer to the talkers, so close they can feel your presence. In this way, you use your body language to make them aware that what they are doing is not what they decided as a team to do.

Often, your physical presence can remind them of their own collective decision on how to spend time during a retrospective. If it still does not work, you need to talk directly with the one who keeps initiating the small talk—preferably outside the retrospective.

If I know a team well, I feel that I am in a position to say, "Remember, this time is set aside for a retrospective. Please respect this time slot and talk about other things at other times." If there is a particular subject the team would like to discuss, you can set up a parking lot where conversations can be parked for later. This solution is generally more useful for the **Overtime** antipattern (Chapter 4) than for this antipattern, though, because here we are dealing with many small conversations that can easily continue over lunch.

ONLINE ASPECT

If the retrospective is done online, you cannot use your body language to make the team stop their small talk, but in my experience, small talk in small groups is not a big problem in online retrospectives. However, if the people in the retrospective are not completely distributed—that is, if some people are sitting together in a room and others are alone in another building or country—the **Small Talk** antipattern is much more likely to happen than it is in an offline retrospective. It is very easy for the people sitting together to start a little conversation about the food they have brought or the meeting they just attended together. In those cases, you have to point out that to ensure everyone is on an equal footing in the retrospective, they should behave as if they were attending an online meeting. You can encourage them to attend the next retrospective alone in a room, as the online participants do.

PERSONAL ANECDOTE

I once gave a course on retrospectives. The attendees were to perform a retrospective, so they had to do the activities and discuss in smaller groups. One of the people was an HR person, and most of the others were developers. I found the HR person to be always talking about unrelated topics when I had started their smaller discussions, and I noticed that she was the initiator of the small talk. Whenever I came to the table where she sat, her conversation had nothing to do with the exercise. The first two times, I pointed to the exercise, smiled, and asked them all to focus.

The third time, I asked her why she was engaging in small talk instead of doing the exercise. She said that, because of her job, she knew about people and she knew they needed to engage in some small talk, that it was important. I told her that might be true, but it depends on the setting, and if you have only an hour to reach some sort of conclusion, then there is no time for small talk. I am not sure she appreciated this, but at least the team was able to continue working on the assignment. Sometimes, you must be a bit stern toward one person in order for the majority to have a better, more useful time.

On the other hand, I have seen some people engage in small talk because they feel they can't contribute anything worthwhile to the discussion. They might be nontechnical yet find themselves in a sea of technical discussions, or perhaps they are new to the team and have very little to add to the discussions yet. This situation might be hard to spot if you yourself are new to the team, but if you know the team members and their situation, you might be able to help the small-talking people by dividing into smaller groups during discussions or even asking them to help you with some parts of the retrospective.

Unfruitful Democracy

... in which, to the frustration of the minority in the team, democracy is used to decide what to discuss and what to do, and the facilitator finds other ways of deciding that makes everyone happier

Chapter 6

CONTEXT

At one of the retrospectives, Sarah is facilitating a discussion about the test framework. Most of the people in the retrospective find this topic very interesting, and when they dot-voted on what to talk about, it was the absolute winner. Unfortunately, Kim and Bo do not feel it applies to them because they do not think that testing is their responsibility. After 2 minutes, they start talking about one of the other issues, one that got fewer votes and was not selected for discussion.

Sarah notices their discussion after a few minutes and starts speaking louder when talking about the next step in the discussion. She believes that even if these people do not seem to care now, the consequences of what the team decides may influence them. At this moment, she does not know how to stop them, and she notices some of the rest of the team seem annoyed with the off-topic talking, while others jump into Kim and Bo's discussion.

GENERAL CONTEXT

At a retrospective, it is often a democratic process that decides what should be discussed. Because the majority always decides what to discuss, some of the issues chosen for discussion are uninteresting to the minority of the people in the room. In their boredom with the discussion, those people start talking about different topics or perhaps different aspects of the same topic. The result is that multiple parallel discussions take place without a plenum follow-up, and some important points are missed by parts of the team, which makes agreement and results impossible. De Bono's *Six Thinking Hats* (1999) really helped me understand how big this problem is and how to avoid it.

Six Thinking Hats

Six Thinking Hats is a system designed by Edward de Bono, which describes a tool for group discussion and individual thinking involving six colored hats. Six Thinking Hats and the associated idea of *parallel thinking* provide a means for groups to plan thinking processes in a detailed and cohesive way and, in doing so, to think together more effectively. The way to create parallel thinking is to show that we are now thinking about the same aspects of a topic at the same time either by expressing that a particular kind of thinking is going on or by actually wearing a hat: a white hat for information, a red hat for feelings, a black hat for critical thinking, a yellow hat for optimism, a green hat for creativity, and a blue hat for structure. That way, we try to avoid having one part of the group thinking about pure data while the other is trying to be critical (De Bono 1999).

ANTIPATTERN SOLUTION

The facilitator is often a very nice person who has empathy for other people. He or she understands that some things are more interesting than others to discuss and that boredom can feel brutal and painful, especially in our modern, boredom-free society. In effect, the facilitator allows people to talk about other subjects in order to keep them happy and in the hope they will rejoin the main discussion when the topic turns to one they are interested in.

CONSEQUENCES

When the team starts to divide their discussions at the retrospective, it is most often because some of them feel they do not need to listen to what is going on at the moment and they can tune in and out as they please, as they do in other meetings.[1]

One consequence is that they might miss information that is relevant to them, even if they didn't expect it to be. When they realize the discussion

---

1. Other meetings should not have divided discussions either, but this is a book about retrospectives.

has become interesting, they ask people to repeat themselves, which is counterproductive for a retrospective or any other kind of meeting.

Another consequence of a divided discussion is that it is disrespectful to the people who are talking. They are clearly talking about the issue because it is important to them and they believe that some change will improve a situation. Ultimately, the retrospectives might be dropped because they do not provide the sharing the team expects and people feel they are wasting their time. **Unfruitful Democracy** can also lead to deeper problems in the team, because a minority of the team always see the topics they are interested in being downvoted and forgotten.

Symptoms

The symptoms of this antipattern can be obvious, since it is hard for a facilitator to ignore people chattering in a small group in the corner of the room. Another symptom is that people ask for recaps of arguments when they suddenly realize that something interesting is going on. Alternatively, if the facilitator asserts his or her authority and prevents the side chatter, some of the team members feel their time is being wasted and that their own issues are never addressed.

Refactored Solution

The antipattern solution of allowing the side chatter is a poor solution. It only scratches the surface of the problem, addressing the symptoms, not the cause. Often, in retrospectives, a democratic process is used to choose which issues to discuss, what experiments to try out, and in what order things are to be done. Dot-voting on Post-it Note issues is a well-known part of facilitating a retrospective.

The trouble with the *democratic* process is that the minority is often over-looked, especially if it is the same people who are asked to vote on the same types of issues at every retrospective. One way of solving this problem could be to use *consensus* decision making, which takes a little longer, because everybody has to agree, but enables the team to shape a decision until a compromise is reached that reasonably satisfies everyone. Unlike the democratic dot-voting, this way of decision making strives to incorporate everyone's perspectives, needs, and, ultimately, permission. The facilitator could also ask for *consent* by everyone, which means that everyone is not necessarily in agreement that this is the best choice but at least that it is the decision that all accept. Of course, the facilitator could also use *dictatorship* in which one person makes the decision because this person knows best or because it will have the most impact on him or her.

Another solution, called *minority vote*, is to separate the votes and display them on different parts of the board, as shown in Figure 6.1. For instance, five of the votes can be used on the 11 subjects from the left whiteboard, and the remaining three votes can be used on the 8 subjects from the right whiteboard. In that way, the facilitator can make sure that the issues posed by the minority will get some attention, even if it is not the focus of the majority. If the minority also get time for their issues, it should be easier to keep them from holding a parallel meeting in the corner of the retrospective.

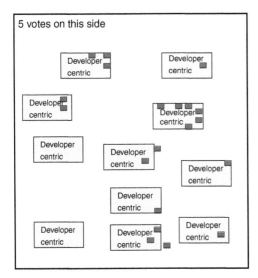 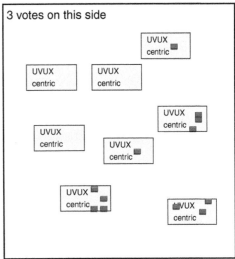

Figure 6.1 Minority vote

From Jutta Eckstein, I learned another option. She divides people into small groups, pairs if necessary, and lets them work on different topics in parallel instead of everybody working on the same topic. They get a timebox for the small-group work, and when it is over, they either come together again and share what they came up with in plenum or discuss another topic in a new timebox. When the topics are discussed and the outcomes from the different groups are shared, all of the team members know the different actions that are proposed and can decide jointly what they want to do. This way, everyone can work on their primary concern and still be informed about what happens in other groups.

It is important to remember that when a team decides to take an action, they must not only believe it is important but also have the energy to do it. Otherwise, even important things are never acted on. I often ask a team to vote for what is important to work on and then, afterwards, ask them what they have the energy or passion to actually do something about. The first time I do this with a team, they often frown and say that passion is less important than impact. But after a few retrospectives, where they have not taken energy into account, they learn that there is a lower likelihood that anything changes without energy.

ONLINE ASPECT

If the retrospective is done online, all the same issues apply. It is a bit harder to change the setting online if this has not been anticipated, but making people discuss in smaller groups is often possible in so-called break-out rooms available in most online meeting tools. If you know that a democratic process will not be optimal for the team, you can prepare online documents to support other types of decision making. For a minority vote, you can create a document divided into sections for the different proposals so that votes can be cast and distributed to the appropriate section. For consensus, I suggest an anonymous vote so that people feel safe to vote freely and are not influenced by everyone else's votes. Consent demands that the vote is not anonymous, on the other hand, since everybody has to give their consent, and the way to achieve consensus is normally to change the suggestion until everybody can accept it. Dictatorship is exactly the same as in an offline retrospective: the suggestions are collected, perhaps voted on, and the person who is to make the decision does so on the basis of the information at hand, the votes, and his or her own knowledge and experience.

PERSONAL ANECDOTE

I once worked at a small Danish company as a retrospective facilitator, and I was facilitating biweekly retrospectives for a team. In an attempt to create a cross-functional, autonomous team, management had included a tester, a UX expert, and a UI expert on the team. The scrum master was also a developer and, thus, we had seven backend developers and three other roles on the team.

In the retrospectives, I noticed a pattern: we often used dot-voting to decide what to discuss or which experiments to implement in the next sprint. This worked really well most of the time for most of the people on the team, but often, the issues that were important to the UX/UI experts or the tester were not voted high enough to even be discussed at the retrospective. In effect, issues such as "introduce UX discussions earlier in the projects" and "spend an insane amount of money on more automated testing" were never voted higher than "better coffee in the lounge" and "a faster server." Consequently, the developers loved the retrospectives, but the rest of the team started to avoid them or just zoned out during them, since the subjects they wanted to discuss were never on the table.

This particular problem was solved with minority voting to make sure that the UX/UI and test issues were also sometimes discussed and solutions sought. Being able to focus on the needs of the minority instead of always supporting the majority is not intuitive, but the effect is dramatic.

Nothing to Talk About

... in which the team believes it has become so good that it doesn't need retrospectives, and the facilitator shows the team how it can learn to keep improving

Chapter 7

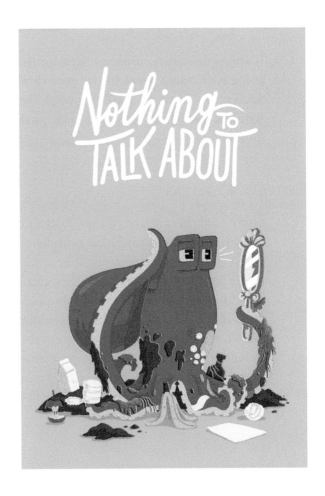

CONTEXT

The team is doing really well. The team members like working in an agile way, and their work product consistently adds value for users. The past two retrospectives have not shown any problems to deal with, and all Post-it Notes have been green except for a few about bad weather (it is Denmark, after all). Sarah overhears Rene and Andrea talking about whether Sarah expects them to invent problems.

It seems there is no longer a need for the retrospectives, since everything is going well. Sarah urges the team to think of issues that need to be addressed, but to no avail. Although she is frustrated, Sarah must agree with the rest of the team that perhaps the retrospectives are no longer useful.

GENERAL CONTEXT

I have seen this pattern of the life cycle of retrospectives in almost every organization and team I have introduced retrospectives to: first, the team might find the retrospectives to be a waste of time—until they learn how helpful and efficient (and fun) they can be. Then comes a period when they find a lot of things to talk about during the retrospectives. It feels good to do something about the problems in the teamwork, the communication, the technology, or the system they work on. After some time, though, the team "runs out of problems." Discussions revolve around the same issues, and the retrospectives become stale and useless. Routine has set in, and the retrospectives become just another meeting that people want to be done with.

ANTIPATTERN SOLUTION

The obvious solution to this is to ask every time a retrospective is planned whether the team members have anything new to talk about, and if the answer is no, the retrospective is dropped.

CONSEQUENCES

To understand the consequences of missing out on retrospectives, first look at the expected benefits: a shared story about what has happened or is happening, a way to let off steam when something is subpar, and a shared decision making about what to do next. The personal factors are also important in retrospectives. This is often when you are all together to have a laugh, share feelings about various issues, and learn how to help each other or yourself. Facilitated in the right way, retrospectives can help you learn and grow as a team. Often, it is the one opportunity for a team to go beyond the obvious and look at the deeper issues. As J. P. Morgan said, there are two reasons for every decision and every action: the official reason and the real reason.

SYMPTOMS

You hear on the grapevine, "We do not need the retrospectives anymore," "There is nothing to talk about," or "We are a good team now—no need to waste time on inventing problems."

REFACTORED SOLUTION

As suggested by the subtitle of Diana Larsen and Esther Derby's book (2006), *Making Good Teams Great*, retrospectives are useful for all teams, not just for teams that are dysfunctional in some way. We can always get better at what we do: we can go from good to great.

World-class skiers still try to improve how they ski: they continually inspect what they do and adapt their technique, their sleep, or their food patterns according to what they find. Even if you go to the dentist to get your teeth cleaned or to a mechanic to get the wheels on your car balanced, you still have to get it done again on a regular basis. This is what I sometimes say to a team that has started the part of the retrospective journey where they believe they no longer need retrospectives.

And then I introduce something new to their retrospectives. I try a new kind of retrospective, such as a *positive retrospective* or a *team radar ret-rospective* (see Figure 7.1), or I make a *futurespective* that will show what they fear or what they hope will happen. As described in the **Negative Team** antipattern (Chapter 21), the team can learn not only from the positive aspects of an issue but also from the negative—which good practices can be made even better.

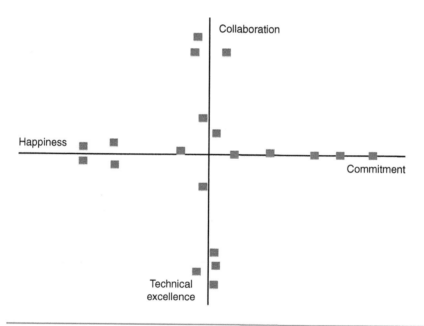

Figure 7.1 The team radar retrospective with four topics

Futurespective

A futurespective is like a retrospective, but the focus is on the future, not on the past. For a futurespective, I build a timeline on the board that starts today (on the day of the futurespective) and goes into the future, perhaps 3 months, perhaps a year, or perhaps until the next release. I then ask the team members to imagine they are in the future, at the end of the timeline. If I dare, I ask them to close their eyes and actually try to imagine what the world looks like "now." Then they open their eyes and add Post-it Notes to the board describing all the positive and negative events that took place from the beginning of the timeline to the end of it.

A typical pushback to this exercise is that people cannot see the future. As a facilitator, you can acknowledge that they cannot predict the future, but explain that what they will be sharing are their hopes and fears for what will happen.

When they have added their expected future positive and negative events, you go through them and discuss what led to them—the causes or the stories behind them. The results of a futurespective can be that the team decides on ground rules that allow them to work in the way they function best, such as no morning meetings. It could also be a note to management about what they need in order to make this team successful—for example, the team should stay together and not be cannibalized by other projects in the organization. In addition, the team might identify experiments, as in a normal retrospective: things they can do for a few weeks and look at the consequences, such as peer review of all work or mob programming once a month.

For me, an added benefit is that I learn a lot about the past of the team members. In their fears, they describe what negative things they might previously have experienced. In their hopes, they describe the essence of how they like to work. Thus, I often have a futurespective with a new team in the beginning of my relation with them.

Another useful exercise is The Ship (Figure 7.2), which can also be a good way to try something new in a retrospective.

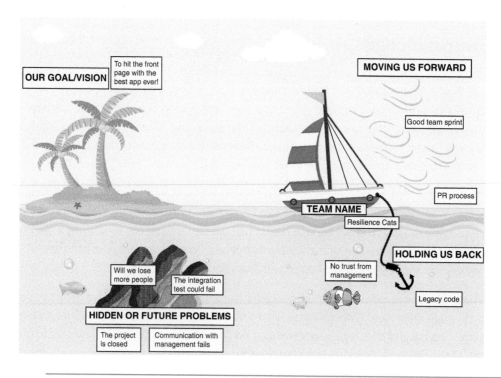

Figure 7.2 The Ship retrospective

The Ship Retrospective

The Ship is a retrospective built around a picture of a ship with a bounty island, wind in the sails, an anchor, and some cliffs hidden under the water. The team must first agree on what is on the bounty island, meaning what is the dream scenario, their vision. Then the ship should be given a name, because deciding on a name for the ship demonstrates how people on the same team can have very different viewpoints regarding the defining factor of that team. I sometimes skip this step for time management reasons.

Now the team members write individual Post-it Notes for the positive things and place them on the wind that enables the sails to make the ship sail toward the island. Next step is the anchor, which represents the things that keep us away from the

(Continued)

island that make progress hard. The last step is the cliffs, which are the things we are unsure about but that we worry might happen. Then the Gather Data phase is done, and the team can start the Generate Insights and Decide What to Do phases. Among other places, The Ship is described in *Getting Value Out of Agile Retrospectives: A Toolbox of Retrospective Exercises* (Gonçalves & Linders 2014).

Another alternative is to ask the team what the worst-case scenario is or how they would work if they had less time, more time, fewer people, more people, and so on. Try to set up another worldview for them and see what they can come up with. Usually, when they cannot see what to change in order to improve, it is because they are so used to the way things are that they have trouble imagining a different situation for the team.

Some teams have learned only to look at their communication, cooperation, and process in the retrospectives, and while these considerations are important, hard data can also be worthwhile to look at in the Gather Data part of the retrospective. Hard data might include burndown charts, regression test overviews, peer review statistics, user feedback, and so on.

Choosing a theme for the retrospective can also be a way to focus and dig deeper into a particular subject. The theme might be a specific release, the test strategy, how we learn as a team and individually, the architecture, or whatever could be interesting for this team.

Sometimes you can allow yourself to play a slightly more active role in the retrospective. For instance, when I hear questions and comments such as "Why don't we just get better at estimating timelines for tasks? Then we don't need to talk so much all the time—we just check in when the work is due to be finished," or "Do we have to discuss the things that go well? Can we not just focus on the things that are bad?" I like to sometimes raise questions about how the team members really want to work. Do they believe in agility? And humanity? And what are their aspirations with this team?

You can also use one of the assessments, such as Agile Fluency, to make the team aware of how they work with respect to agile methods, standards for code quality, or team happiness. Then you can start a discussion about where they want to be or where they thought they were.

Agile Fluency

Learning to work in an agile way is like learning a new language. You may learn to speak the language, but when something unexpected happens, you switch back to your mother tongue. For instance, I can speak German, but when I stub my toe on a table leg, I swear in Danish. If I am not fluent in a language, I switch to a language I am fluent in when I am surprised, angry, or afraid. The same goes for agile work in teams. They might have taken some courses and learned how to work with Scrum, but when something unexpected happens, they switch to waterfall because that is what they know and understand. The point of Agile Fluency is to assess and visualize what level of fluence a team has with agile methods.

Diana Larsen and James Shore cofounded the Agile Fluency Project in which they help coaches enable teams to work at the level of agile fluency that best fits their business's needs.

The point of this refactored solution is to support them in finding ways to improve every aspect of the team and their work, if only in small ways.

ONLINE ASPECT

If the retrospective is done online, all the same points apply. The only difference is that it is harder to change the agenda of an online retrospective, so you might not be able to change this antipattern when you encounter it, but you will be able to plan for it for the next retrospective.

PERSONAL ANECDOTE

As mentioned previously, I have seen this particular part of the retrospectives life cycle with numerous teams. With one particular team, it was hard for me as well to see how they could improve. They seemed to be doing everything just right and were very nice to each other while doing it.

I decided to prepare a retrospective around the word *brave*. The Set the Stage phase included a question on how each team member had been brave in the last month. I asked for just one example, and it could be at work or in private. (Naturally, we also went through the action points from the last retrospective and discussed whether the experiments had been successful, what they had learned, and what they would continue doing.)

In the Gather Data phase, I asked the team members to think about three things they would do if they woke up one morning and were ten times braver than usual. At least two of these items should be related to their work. You might argue that all three items should be related to their work, but I chose to include a possible private setting for two reasons.

First, this question can lead to some fun facts about each other, things they have always wanted to try but never dared. Fun facts can give the team a laugh and create a stronger bond among the group. Sometimes, the team members even find out they can help each other achieve a "brave goal." One time, for instance, when someone said that he wanted to learn how to fly, another team member invited him to be her copilot next time she went flying.

Second, by leaving room in the question for people to think outside of work, you may encourage them to talk about an important personal wish that they wouldn't ordinarily talk about at work but that nevertheless preoccupies their thoughts. When they share their important dreams or goals instead of trying to stifle their thoughts, they are less distracted and better able to focus on their work.

What came out of this exercise were answers as different as "talk directly to future users of the system in the nearest mall," "try mob programming,"[1] "ask for help when I run into something I cannot understand," and "reeducate myself into a backend developer." The next phase, Generate Insights, was about the stories behind their answers. If people wanted to, they could share the experiences that led to their list of what they would do if they were ten times braver. Had they already tried and failed? What happened when they tried? What made them try, and what did they expect to gain?

For the Decide What to Do phase, each person chose one idea to focus on or to help someone else focus on. And then I helped them figure out what a less frightening first step could be. The retrospective brought some new issues to the surface, because the team members were forced to think in a different way. It was easier for some than for others, but everyone in the room had some reflections about their normal behavior and whether something could be improved on a personal level.

Another aspect of this antipattern is that a team thinks right from the outset that retrospectives would be without value for them for a certain reason. I recently had somebody tell me that "agile coaches and scrum masters are getting nervous, because DevOps is happening everywhere around them. If scrum masters and agile coaches knew about DevOps, their whole perspective on retrospectives would change. DevOps is about measuring outcomes. In that case, the retro should focus on those outcomes, and many DevOps outcomes are technical: production incident rate, feature cycle time, mean time to recover/restore service. You can't facilitate discussions about those things without knowing some DevOps."

I really like statements like this one because it gives me a chance to explain why I think retrospectives are valuable. At least the person who made this statement will not implement retrospectives just for the sake of retrospec-

---

1. Like pair programming but with the whole team programming together following the rules written by Woody Zuill.

tives. I replied to this comment that, first, even if people are using DevOps, they are still people, and they still have all the same people problems as everyone else.

Second, even though I know most people think about data gathering in a retrospective as gathering data about incidents and feelings, I also often see burndown charts, regression test results, recovery time, and reply time. I sometimes facilitate retrospectives where the team has decided to look only at technical subjects, which can work really well. Often, though, we talk about people issues as well, because of questions such as "How did we end with that response time?" or "Why did we get that regression test result?"

Third, I know I benefit from my technical background, and when people dive into technical discussions, I am able to shortcut them sometimes because I can understand what they say and know when they start repeating themselves. But many facilitators do not have that background and are still perfectly able to facilitate retrospectives. We, as facilitators, are supposed to be invisible and supportive of the discussions, and with the right amount of understanding of body language and team dynamics, it is still possible to facilitate a meeting about something you know nothing about.

Fourth, and this is specific to DevOps, I currently facilitate retrospectives for two teams, an operations team and a development team, who are in the process of being merged into a DevOps team. They have specific issues relating to their separate teams being merged into one. The dream, of course, is that they work as one team and that developers have not only sympathy for the operations but also empathy because they will learn what is done in operations—and vice versa for the people in operations. This poses some challenges when it comes to respecting each other's knowledge and skill sets as well as creating a learning environment in which they learn from each other so they can make it possible to collect data and react to that data. There is a lot of tension between these two teams, and the retrospectives are helping them respect each other, albeit slowly.

Political Vote

. . . in which the team members wait until the last moment to vote in order to game the system, and the facilitator finds a way to make the voting system more fair

Chapter 8

CONTEXT

Sarah wants to use the retrospective as an opportunity to discuss how the team values the quality of their work. This subject is somewhat delicate, since some of the team members believe that the quality is very low. Sarah knows this because she has talked with some of them offline, but she also knows they never discuss the work quality openly and that the tension around it simmers in the background of each meeting.

This retrospective has to be done online, since some of the team members are working remotely this week.

She decides to use a team radar retrospective (Figure 8.1), also known as a spider web retrospective. She has prepared a Google Drawings diagram and labeled the six spokes Quality, Customer Value, Test Coverage, Internal Communication, External Communication, and Fun.

Using this approach, Sarah can evaluate the team's thoughts on quality without making it too obvious, because other topics are represented on the spokes as well. The task for the team members is to place a virtual Post-it Note on every spoke, rating these six aspects of the team on a scale from 1 (poor) to 5 (excellent), to visually show the group's collective opinion.

Had it been a retrospective with everyone physically present, Sarah would have asked people to write the numbers on a piece of paper, and she would have read them aloud to protect anonymity. Then everyone would have looked at the votes and discussed whether the results were surprising, whether they were good or bad, and what could be done about them. Since this is an online retrospective, she needs to use online tools, in this case, Google Drawings, which allows people to place their virtual Post-it Notes anonymously. The downside to this voting method is that everyone can see the votes as they are placed and may be influenced by those votes. For example, they might see that the first two votes rated quality as excellent, so to avoid being the reason for discussing quality, they decide to follow the votes of the first two.

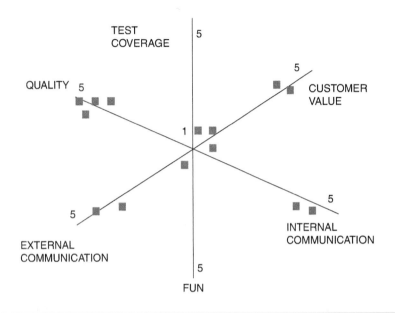

Figure 8.1 A team radar retrospective with six topics

GENERAL CONTEXT

In a retrospective in real life, it is often harder to make things anonymous because we can see where people put their Post-it Notes or we know their handwriting. In an online retrospective, it is easier to maintain anonymity when needed. But in the process, we must make sure that we do not lose other things, such as timing, and that everyone votes without looking at other people's votes.

ANTIPATTERN SOLUTION

The antipattern solution is to let people vote by dragging and dropping virtual Post-it Notes to the place in the image representing their choice or opinion. The timing in this approach can be a problem, though, because people may wait until they see what other people vote for before they cast their own vote. Also, in contrast to an in-person retrospective where you might walk around the room and collect a piece of paper from everyone, it can be difficult in an online retrospective to see whether everyone has voted.

CONSEQUENCES

When people can see other people's votes before they themselves vote, they sometimes cast a "political" vote, or a politically correct vote, rather than one that reflects their true opinion. Perhaps they do not feel safe in this team, so even though the votes are anonymous, they do not want their vote to lead to a discussion in which they are obliged to express aloud an opinion that disagrees with that of the majority. This phenomenon might influence everybody to vote in approximately the same way. Consequently, the issue remains an unacknowledged difference of opinion and a source of tension within the team.

SYMPTOMS

The symptoms of the **Political Vote** antipattern often are easy to spot, if someone says out loud, "I will wait until everyone else votes, and then I'll vote where it makes the most difference according to what I want to talk about." If two or more topics were tied, for example, this person could sway the outcome by casting the tiebreaking vote for the topic he or she prefers (even if it wasn't the person's first choice). Other people might withhold their vote until they see what others vote for, then vote with the crowd—that is, cast a politically correct vote, as described previously.

REFACTORED SOLUTION

The refactored solution is to have all the team members vote at the same time. Most online tools support simultaneous voting: you ask everyone to "pick up" their vote, hold it for a count of "1-2-3-vote!" and then drop it into place. When everyone votes at the same time, you get a truer picture of what the majority wants. If your online tool does not support this feature, you can ask people to send you their votes in a private chat, and then you can place all the votes in the radar. Some online retrospective systems, such as Retrium, have voting systems that do not allow you to see the votes of other people before you have voted.

ONLINE ASPECT

The context of this antipattern is an online retrospective, but in an offline retrospective, you could solve it by having everybody hand their votes to you, and then you put all the votes on the board yourself.

PERSONAL ANECDOTE

I was facilitating a retrospective in a team I had been working with for a long time. We knew each other. Our retrospectives were always online and always with the use of Google Drawings as a canvas for the retrospective. This time we held a team radar retrospective, and we wanted to figure out what aspects of their work they wanted to focus on.

The problem was that one of the team members did not want to vote until everybody else had voted. He felt insecure about evaluating aspects of work such as code quality and test coverage because he was worried that his opinion might be unpopular. Even though it was a completely anonymous vote, he knew that if he rated, say, code quality as 1 or 2 when the rest of the team rated it 4 or 5, the ensuing discussion would require him to support and defend his unpopular opinion.

From my point of view, I needed his input and valued his opinion—and so did everybody else on the team. We resolved the problem by using the 1-2-3-vote method every time the team had to vote. When everyone voted simultaneously, no one could base their vote on other people's votes.

Planning Antipatterns

Team, Really? ...in which the borders of the team are blurred, and the team members all help each other figure out who should attend the retrospective

Do It Yourself ...in which the facilitator is wearing several hats, which is suboptimal for both the facilitator and the retrospective, and the team finds other facilitators to take over at times

Death by Postponement ...in which the team is so busy with "real work" that the retrospectives are postponed again and again, and the facilitator helps the team see how valuable these retrospectives are and that they are real work

Get It Over With ...in which the facilitator rushes through the retrospective in order to "waste" as little time as possible for the team, and the facilitator finally decides that to have a decent retrospective, sufficient time must be allowed for discussions

Part II

Disregard for Preparation ...in which the facilitator initially misjudges how much preparation an online retrospective requires and later learns how to prepare for it wisely

Suffocating ...in which team members get tired and hungry and unfocused during the retrospective, and the facilitator makes sure to feed them and give them oxygen so that they can concentrate a bit more

Curious Manager ...in which a manager is curious about what happens at the retrospectives and wants to listen in on them, and the facilitator, in a nice but firm way, says no to the manager

Peek-A-Boo ...in which team members will not show their faces on the video in an online retrospective, and the facilitator learns why and finds ways to make it safer for people to show their faces

Team, Really?

. . . in which the borders of the team are blurred, and the team members all help each other figure out who should attend the retrospective

Chapter 9

CONTEXT

The team lacks sufficient software architecture knowledge, so a software architect is assigned to help in this area. Bo is not happy about it, since he believes himself to be a software architect and that the team therefore needs no outside help. Bo does not say so out loud, though, since that would be admitting that the team has not acknowledged it to be true. The architect is assigned not only to this team but also to three others, and thus is not with any team on a full-time basis.

On first consideration, it sounds like the architect will work 25% of the time with each of the four teams, but whenever we share our focus between two or more things, we lose a percentage of our time due to context switching.[1] How much time we lose depends on a number of circumstances, but a good estimate is between 10% and 40%, according to research. To reduce the amount of time lost to context switching, the architect decides to spend two consecutive days with each team instead of moving between teams every time one has a question for her to answer or a meeting they want her to attend.

It turned out the teams had a lot of questions for the architect and started following her around. Whenever she went for a break, people would line up by the coffee machine or even outside the restroom and inundate her with questions. Finally, the architect started bringing her own coffee and using a different restroom.

The effect is that whenever a team has a question for the architect, they have to wait up to 6 days for an answer because the architect is working with other teams. When weekends, sick days, and travel are factored in, it can take even longer. It is not a surprise, then, that the teams often must find answers for themselves, and Bo is happy again. The system suffers, though. Bo's decisions sometimes have ripple effects on parts of the system other than the one he and his team worked on. They are not problems

---

1. https://productivityreport.org/2016/02/22/how-much-time-do-we-lose-task-switching

that showed up in the regression tests but subtle changes caused by disregarding the architect's guidelines. Each problem generated a lot of shaming and blaming between the architect, the other teams, and the team in question, most of it directed at Bo.

GENERAL CONTEXT

When deciding whom to invite to a retrospective, many things should be taken into consideration. Should the manager be in the room (see Chapter 15, **Curious Manager**)? Should the student intern be present? Should the experts who assist us from time to time be present? And should they be there for all of the retrospectives or just select ones? Often, the people who are working with different teams want to optimize their time spent with the teams, so they skip social time, most standup meetings, and retrospectives.

ANTIPATTERN SOLUTION

The solution we often use is to save time for the experts by not inviting them to the retrospectives.

CONSEQUENCES

A retrospective where an important part of the puzzle is missing is subop-timal. One of the main outcomes of a retrospective is that the team shares how this last sprint/week/project/year was for them: what worked, what made them happy, what gave them energy, but also what actually hap-pened. I often hear one or more team members say, "I did not know this!" while gathering data. Sometimes people expect everyone else to know what they themselves know, but even when people work together in an open office, there are things, important things, that some people miss.

The picture that is drawn by gathering data from the whole team is invalu-able in understanding, or in other words, inspecting, how the work is done in the team. The experiments that are decided on—the changes made to the way the team works together or with the technology—are also important outcomes of a retrospective because these experiments are the way the team adapts to the situation at hand. When a member of the team or someone who works closely with the team is absent from the retrospective, the actions decided on may not be acceptable for all. Perhaps the decisions even cause problems, such as with the architecture, the UI, or the tests, that the missing team member could have foreseen. This *inspecting* and *adapting* is the heart of agile development.

Last but not least, I have often heard people say that a retrospective feels like group therapy for the team and that it is where they start sharing and feeling like part of a team. People who do not attend the retrospectives consequently might feel less like part of the team.

SYMPTOMS

Topics introduced at the standup meeting are important for the work of someone who is not there. A person crucial for implementing a user story has not been involved in creating it. Decisions are made on a daily basis without consulting someone whose opinion should be taken into account.

REFACTORED SOLUTION

Everyone has to attend the team retrospectives. When I say everyone, I do not actually mean "everyone"—I mean the core team and the people needed to create a reflection on the previous way of working and to make decisions about how to work together in the future. The next question is, How do we know who is needed?

The manager might be needed if the team has some issues that are outside their circle of influence (see Chapter 3, **In the Soup**) or if they want to address something they need the manager to hear about. The same is true when it comes to people such as architects, UX experts, and testers who have specialist skills that are needed in many teams. If you are working on parts of the system that will change the architecture, or if the architecture is changing and has a consequence for your work, you should invite the architect. The same logic applies for testers and other experts. You could argue that everybody should be able to test, for example, but the reality is often that even if the developers know how to test, they need a test expert to make a plan and set some boundaries for the test strategy. These aspects of the work are so important that it is my belief that the experts need to be at all retrospectives in order to be there when information is exchanged about how it is to actually work within the boundaries they have described.

The architects need to be present when decisions are made about how to work with the architecture (or around it). I would also argue that for someone who is on the periphery of the team, it is even more important to be there when decisions are made about how to work with each other in the team.

Some solutions to the absence of experts at retrospectives, of course, are not even related to retrospectives. Perhaps the company needs to hire or train more people with the needed skills. Perhaps the way of working could be made more convenient to accommodate the circumstances, such as having a kickoff meeting at the beginning of each project to make decisions on how to communicate with the people they need.

ONLINE ASPECT

Since this antipattern is about whom to invite to a retrospective, an online retrospective has the same issues as an offline retrospective. One small difference is that the people invited to the retrospective from outside the core team find it more convenient to attend because they can be there remotely and because the online retrospectives are often shorter than the offline ones. I have also facilitated retrospectives where the team invited the manager in for just part of the retrospective, which is also easier online.

PERSONAL ANECDOTE

This anecdote is from my work as an agile coach at a large Danish company. Four different teams worked on different parts of a product, and the teams were cross-functional with all the expertise and skills needed in every group. Unfortunately, some of the expertise was to be found in only a very few people at the company, so these experts (e.g., testers, UX experts, and architects) had to be part of several different teams at the same time. They divided their time between three or four different teams, and in effect, they had to rely on each team to follow their guidelines even when they could not be there themselves to ensure their decisions were adhered to.

Because of the enormous load on each of them, being shared between many teams, they wanted to spend time on only the most necessary meetings. Retrospectives have a tendency to be valued less than other meetings, and as a consequence, the experts chose to avoid them. At one retrospective, there was a discussion about the UX side of the product, and everyone had an opinion. A decision was made to go back to the earlier UX design, since this would speed things up development-wise for this particular part of the system. The team took the action point from the retrospective room on a Post-it Note, and the development continued with the earlier UX design.

After about a week, the UX person had time for that team again and attended a standup meeting in the morning. At this standup, people shared what they had done and what the obstacles were, and the UX person understood that the team had been aligning their work with an earlier version of the UX design. A discussion followed, quite heated, and it turned out that there was a very good reason for abandoning the earlier UX design. Of course.

Do It Yourself

. . . in which the facilitator is wearing several hats, which is suboptimal for both the facilitator and the retrospective, and the team finds other facilitators to take over at times

Chapter 10

CONTEXT

Sarah feels responsible for keeping the team on the true path of agility. She has to crack the whip to get people to the daily standups, and every 3 weeks, at the end of the sprint, she makes sure a retrospective takes place. It is natural that Sarah, the scrum master, facilitates the retrospectives—*all* the retrospectives from first to last.

In the beginning, the team members found every retrospective to be interesting, useful, and sometimes even fun. Time has passed, however, and the team is getting tired of the way the retrospectives are done. They feel the retrospectives waste their time because they are stale and unproductive. They blame Sarah for being a bad facilitator and allowing the retrospectives to become boring and inefficient. Sarah is also becoming frustrated. She feels that she does not get anything out of the retrospectives. In addition, she has to be above the fray—careful not to become directly involved in any of the team's tense or angry disagreements during the retrospectives, and she feels less and less like a part of the team.

How can Sarah reflect on her own issues when she is always occupied with facilitating?

GENERAL CONTEXT

An important part of being agile is to inspect and adapt. Having regular retrospectives is an obvious way to achieve this, and someone has to facilitate them. This "someone" often turns out to be the scrum master, which is not in itself a bad idea, but it can have negative consequences.

ANTIPATTERN SOLUTION

Who should facilitate the retrospectives? For various reasons, the answer is often the scrum master. Scrum masters might appoint themselves because they believe (perhaps rightly) they would be best at it. They might be appointed by management, which believes facilitating is a scrum master task. In many cases, no one else dares to, or wants to, try their hand at facilitating.

Many articles and blogs describe the facilitation of retrospectives as the scrum master's responsibility, and many companies I have worked with have decided that the scrum master is the obvious choice for facilitation of retrospectives.

CONSEQUENCES

There are two alternative negative consequences of this antipattern solution. For both, the essence of the problem is that you cannot wear two hats in a retrospective. You cannot be both facilitator of and participant in a retrospective. Both roles require your full attention.

Scrum masters are so much a part of the team (hopefully) that they get deeply involved in the activities and discussions. When a scrum master also serves as the facilitator, he or she still wants to participate as a team member in order to benefit from the retrospective. The consequence is that he or she forgets to facilitate—to keep track of time and to be aware of the energy in the room (see Chapter 14, **Suffocating**). In addition, the scrum master may have a hard time being objective when it comes to changing the agenda of the retrospective in order to adapt to the situation.

The alternative drawback is that the facilitator is so eager to fulfill the role that he or she completely forgets him- or herself in the polls, brainstorming, and so on. Thus, the scrum master's voice as part of the team is not heard.

SYMPTOMS

You may discover that you are in this antipattern if the facilitator is unaware of the time or the energy in the room, or if the scrum master feels unfulfilled after retrospectives. You hear the team members making comments such as "The retrospectives are boring," "They are a waste of time," and "We should have a better facilitator," and the facilitator even begins to complain: "I would like to get something out of the retrospectives as well." Some of these comments could be rooted in other reasons, of course, and the following solutions might also remedy these causes.

REFACTORED SOLUTION

One solution is to take turns in the team facilitating retrospectives so that everybody gets something out of retrospectives. Being the facilitator once in a while means you will, on those occasions, get less out of the retrospective as a team member. This can be a good solution if there are several people available who are able to facilitate retrospectives. As we have seen, not everyone has the skills, training, and experience, but the skills can be taught. I often teach members of an organization to facilitate retrospectives before I leave the organization (as consultants must at some time). I start with where the energy is, and it might be found in only one or two people in the beginning, but when others see how much the first team members enjoy the role, I normally get a second wave of people who want to learn. I start with a course to teach them the fundamentals, then I let them observe me, and then I observe them and give them feedback. After that, they are ready to facilitate on their own.

It is said that introverted programmers will stare at their shoes when they talk to you. In contrast, extroverted programmers will stare at your shoes. Notwithstanding the value of occasionally checking the shoes of participants, if your team consists mainly of such people, it may be hard to rotate the facilitating role. There is help out there, though, for those who would like to work on their facilitation skills, and I can highly recommend *Facilitating with Ease!* (Bens 2005), *Facilitator's Guide to Participatory Decision-Making* (Kaner 2007), *The Skilled Facilitator* (Schwarz 2002), and *Collaboration Explained* (Tabaka 2006).

Also, some tools[1] exist that make facilitating a retrospective easier. For example, Retrium is an online tool that allows you to choose a specific kind of retrospective and then takes you through everything you need to do in every phase of the retrospective. You could also sign up for a Recesskit, which means that you will receive a package every month with everything you need to facilitate a fun and interesting retrospective. Another tool is Retromat, an online tool that decides on five activities for you, one for each phase, so that you can be sure to be inspired with new activities. My personal favorite is Dialogue Sheets, made by Allan Kelly, which is like a board game you can print out and use to create the dialogues needed for a retrospective as well as other kinds of meetings.

If you rotate the facilitator role, an added benefit is that the fresh approach of each facilitator often adds new energy to the retrospectives.

I often find that if other members of the team try facilitating, they learn how difficult it is to do it right and consequently have greater appreciation for a well-facilitated retrospective. It may look easy to facilitate retrospectives, but it is not—at least not if you want to do efficient, productive ones.

Another solution is to get someone from another team in the same company to facilitate. Or you might even consider hiring a complete outsider, a professional facilitator. Since I have personally facilitated many retrospectives as an external consultant, brought in for that express purpose, I am somewhat biased in favor of this solution. It does add an extra expense to get an external consultant to facilitate the retrospectives, but that cost can often be negligible when compared to the cost of wasting an hour and a half of the developers' time in a retrospective that is not facilitated well.

---

1. https://www.retrium.com; https://recesskit.com; retromat.org; allankellyassociates.co.uk/dialogue-sheets

ONLINE ASPECT

Since this antipattern is about deciding who should be the facilitator, it applies equally in both on-site and online retrospectives. But since the refactored solution often involves asking a less-experienced facilitator to facilitate, there are some subtle differences. In some ways, an online retrospective is easier to facilitate because of the online tools (e.g., Retrium) that guide you through a retrospective one step at a time. Also, the huge benefit of an experienced facilitator's ability to read body language is lost in an online retrospective, so beginning facilitators are apt to do nearly as well as experienced facilitators in an online meeting.

PERSONAL ANECDOTE

In a company I worked for, I turned out to be the one who wanted to—and was able to—facilitate retrospectives. I did it for my own team, and I started doing it for other teams as well. Even though I love facilitating retrospectives, this took time away from other work I also needed to do. In addition, as mentioned earlier, it robbed me of the chance to enjoy retrospectives with my own team.

We decided to make a shared document (see Figure 10.1) with all the teams that needed facilitation and all the facilitators. As a part of this plan, I had to mentor some of them into becoming good facilitators, of course. When a team had set a date for a retrospective, they would then book one of the facilitators. In that way, the facilitator could be someone from another team, and pooling the facilitators had several benefits.

First, all team members, including the scrum master, were able to enjoy the shared learning of the retrospective with their team and to take part in deciding what to try out during the next sprint.

Second, the facilitators could focus on facilitating and not be so tied up in the discussions that they couldn't keep track of time, energy, body language, and all the other important parts of the task of facilitating.

Team/Date	June 1	June 15	June 30	July 15	July 30	August 15	August 30	Sep 15	Sep 30
TheFabulous4	Aino		Aino		Aino		Andrea		Aino
UnicornswithRainbows		Rene			Rene			Rene	
FooBar	Kim		Kim		Kim		Kim		Kim
MemeGenerators	Rene		Aino		Rene		Aino		Rene
GrumpyOldMen	Bo				Bo				Bo
Facilitators									
Aino								X	
Rene			X	.	X	X			
Kim									
Bo									
Andrea	X		X	X					

Figure 10.1 Facilitator rotation

Third, sometimes it can be good for a team to try out new facilitators, since facilitators all have their own style and energy level. In this company, some teams always wanted the same facilitator, other teams wanted a different facilitator for each retrospective, and some had no strong preference.

Fourth, attending retrospectives of other teams can provide ideas for your own team (e.g., mob programming, having meetings only in the afternoons). One of the rules of a retrospective is that there must be trust that nothing shared in the room ever leaves the room unless everybody agrees to it. This is also known as the Vegas rule: What happens in Las Vegas stays in Las Vegas. In light of that rule, I do not mean that you should run back to your team and tell them what the other team is struggling with, but if the team you facilitate comes up with an interesting experiment for the testing procedure or a fun way of making peer review, then perhaps you can "borrow" it.

A related antipattern is **Curious Manager** (Chapter 15), since managers sometimes propose themselves as retrospective facilitators to "figure out what is going on" or even to tell the team what to do. As discussed in that antipattern, a manager as facilitator is not always beneficial.

Death by Postponement

. . . in which the team is so busy with "real work" that the retrospectives are postponed again and again, and the facilitator helps the team see how valuable these retrospectives are and that they are real work

Chapter 11

CONTEXT

In our little team, some problems arise. Some of the code committed to the version control system is not in good enough shape to be committed to the rest of the system. Later, tests and integrations fail and changes have to be rolled back, which means a setback not just for that code but also for the code that depends on it. Frustration starts to build up, and people are looking at the causes for the bad code. A few people start thinking that one of the team members, in this case, Peter, should work on simpler or unrelated problems. Team members become increasingly frustrated, but since they know there is an upcoming retrospective where they will be able to discuss the problems, they wait and abide.

At the next retrospective, the room almost explodes with blaming, and a lot of aggression is vented. The retrospective itself is immediately a success, since Sarah is able to divert the blaming into a constructive discussion of code quality and what it means for the system. The team comes up with experiments to make in the next sprint to achieve better code quality. Among these are standards and pair programming. But at this retrospective, no new problems are discovered and no news is shared. It is merely a focused discussion for that particular problem.

GENERAL CONTEXT

The team has learned that the retrospective is the venue for sharing incidents. They are happy with this arrangement, since it means they have a structured way to discuss issues. Also, many of the team members would rather not talk about things that go wrong, so postponing those discussions feels good.

ANTIPATTERN SOLUTION

After a period of time with regular retrospectives, teams tend to accept that retrospectives are now the setting in which to discuss problems, and so they do not want to "waste" time discussing problems during sprints.

When something happens that angers or frustrates them, they simply try to forget about it until the next retrospective.

CONSEQUENCES

There are two major negative consequences of this antipattern solution. One is that solutions to problems are delayed if you have to wait for up to 3 weeks to solve them, and the issues that are arising may be symptoms of a larger problem that needs to be addressed right away. It could be that someone is getting too stressed to do his or her job properly or that the team is working in the wrong direction. Precious time is wasted waiting for the right time to address these issues.

The second problem is that if you postpone everything until the retrospective, you could end up with too many problems at the retrospective and consequently not have enough time to explore what was not known at the start. Some of the beauty and power in retrospectives come from seeing the whole picture or seeing it from someone else's viewpoint. The sharing of knowledge and reactions to events are as important as the experiments you choose to make on the basis of the knowledge.

SYMPTOMS

If the data gathering feels like an explosion with a single, or very few, subjects, this is a sign that we have entered **Death by Postponement**. If you work closely with the team during the sprint you may also overhear comments such as "Let's discuss this at the retrospective" or "Here is a thing for the retrospective." At the retrospective, you might notice that some issues are not being addressed because there are too many issues.

This antipattern is in itself a symptom of a problem: if a team waits for a facilitator to plan a retrospective before they start discussing important issues, then they are not a self-organized team. So, when you notice this antipattern as a facilitator, you might want to work on this issue with the team, even if they are not aware of it themselves.

REFACTORED SOLUTION

A retrospective should happen every 2 weeks. When the time between the retrospectives is longer than that, it becomes increasingly hard to remember what happened and what you felt about it. Some teams, especially the ones that practice mob programming, have a retrospective each morning to see what they can learn from the day before. In those cases, the retrospective is naturally very short.

Some teams have retrospectives every 3 or even 4 weeks because that is the amount of time they choose to set aside for learning. Then the time between a problem occurring and a solution being found can get very long. Instead of waiting for a retrospective to discuss the issues that arise, address the problem when it occurs. You could, for instance, use a *continuous retrospective timeline*, as shown in Figure 11.1. This solution was described to me by Linda Rising, only then she called them *real-time retrospectives*.

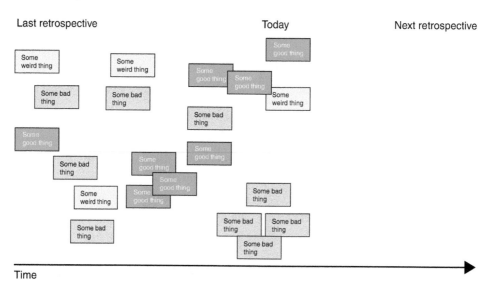

Figure 11.1 Continuous retrospective timeline

This is an implementation of the retrospective activity Timeline, described in *Agile Retrospectives* (Larsen & Derby 2006). Since it is most effective if everyone is able to view it every day, it should be placed in the project room if possible. If not, then it could also be posted online. With a continuous retrospective timeline in the room, team members experiencing something that makes them happy or sad or angry can put up a Post-it Note when it occurs. In that way, you have a shared vision of what is going on with people during the sprint. It should be possible to add Post-it Notes anonymously, and someone should be responsible for keeping the timeline free of blaming, much like a moderator on social networks.

When a bunch of red notes accumulate together in a continuous retrospective timeline, it may be time for a conversation regardless of when the next retrospective is scheduled. In setting up the timeline, it is understood that anyone can call an immediate meeting or even a retrospective if he or she thinks the problem needs a more time-structured discussion. This solution has two benefits: (1) the problem is handled in a timely manner, and (2) there is time in the regularly scheduled retrospective for new issues to be dealt with.

I know companies that use this real-time timeline and are able to shorten the next retrospective because problems were addressed in an impromptu retrospective during the sprint. Most of the time, however, the retrospective proceeds as planned, since we often need some time to get into the right mindset to get something out of a retrospective.

ONLINE ASPECT

For an online retrospective, I often set up the online document that I plan to use in the Gather Data phase right after the last retrospective. In that way, the team members can add items to the document during the time between the retrospectives, and everybody can keep an eye on what is happening. It is my experience, though, that you have to remind the team of the online document from time to time so that they remember to add items.

PERSONAL ANECDOTE

When I entered the room, I could immediately feel the elephant. This was a team for which I had facilitated retrospectives for about a year. Although it had not been so in the beginning, they now looked forward to retrospectives and saw them as an opportunity to vent, to share, to laugh, and to come up with experiments that would slowly but gently turn this great team into an awesome team. They were, in other words, really on board with the retrospectives.

We started the retrospective as usual, with a round of comments related to the sprint (sometimes the comments are unrelated, but that is not the point here), then we went through the experiments decided on in the last retrospective to share the status and find out if the experiments had achieved the expected result.

Sometimes this opening takes a long time, because people have experienced events in different ways, but this time it went really quickly. Two of three experiments had been performed; there had not been a chance to do the third, and I made a note of that in my little black book. Then it was time to gather data since the last retrospective. We used the Timeline, which is one of my personal favorites and also the one that this particular team wanted to use every third retrospective. They liked to try new things, but they also appreciated knowing some things in advance.

The timeline was the first place where I noticed the contours of the elephant in the room. Actually, looking at the timeline, everything was somehow connected to this elephant. Everybody had the recent change in management as a focus area. It turned out that the boss of their boss had quit the job (or had been fired—they did not know) a week ago, and they were all worried how this would affect them.

The last time this had happened was no more than 6 months earlier, and it had had really unpleasant effects on the team. Their immediate leader had been let go, and one of the team members had been transferred to another branch.

It was great that the team—all of them—wanted to talk about this situation, and it was great that they felt safe enough to share their worries. What I was not so happy about was that they had not felt they could talk about the change in management and possible ramifications without a retrospective.

I am always hoping that regular retrospectives strengthen the reflection muscle and that people start reflecting on a daily basis and changing what needs to be changed. To me, it is like waiting until Monday to start a diet: postponing does not improve the experience or the effectiveness.

Get It Over With

... in which the facilitator rushes through the retrospective in order to "waste" as little time as possible for the team, and the facilitator finally decides that to have a decent retrospective, sufficient time must be allowed for discussions

Chapter 12

CONTEXT

All the different antipatterns that the team experienced made people wary of the retrospectives. Even Sarah lost her enthusiasm because of negative feedback and the uncomfortable situations that prior antipatterns put her in. Therefore, when the developers asked to shorten retrospectives from 1.5 hours to 1 hour to give the team more time for "real work," she agreed.

After a few months, it was decided that half an hour would be enough time, and slowly but surely, the retrospectives became nothing more than standup meetings where people were paradoxically allowed to sit down. They would go into the same room (those who did not have better things to do) and take a round to say how things were. On those days, they would skip the morning standup meeting because they already had a status meeting. This antipattern is also described as the *Rushed Retrospective* by Stefan Wolpers (2017) in his blog post about sprint retrospective antipatterns.

GENERAL CONTEXT

As developers, we are often taught that the only time we are creating value is when we are writing lines of code. Thus, time for coding is more important than for other tasks, and time spent on retrospectives seems to be wasted. This can become a self-fulfilling prophecy if people spend so little time on retrospectives that they get very little out of them. Bad retrospectives are a waste of time, as discussed in Chapter 1, **Wheel of Fortune**, and sometimes having a bad retrospective can be worse than doing nothing, as described in Chapter 2, **Prime Directive Ignorance**.

ANTIPATTERN SOLUTION

If the retrospectives start to go stale, the team often decides to spend less time on them, and this shaving off of time continues until the retrospective has turned into a complaint session[1] or a status meeting. In the end, the retrospectives vanish and the chance to learn and adapt diminishes. Also, the sharing of stories becomes more difficult, and the improvements that retrospectives promise can now happen only with a team that is communicating often and working closely together either in real life or with support for distributed work.

CONSEQUENCES

The actual time spent on retrospectives is saved, but the time savings is offset by errors and miscommunications that could have been avoided through retrospectives. Additional time is lost when people leave, taking their knowledge and skills with them, because they are dissatisfied with the work environment or the communication within the team.

SYMPTOMS

You hear people saying, "Let's skip this retrospective. We are busy now—we can have one after the delivery," or "Couldn't we do this in a shorter time? How much difference can 15 minutes less make?" You may even notice that people attending the retrospectives are working on their computer or focusing on their phone, just waiting for something that is important to them.

REFACTORED SOLUTION

How do you resolve this antipattern solution, that the retrospectives are unpopular and are slowly disappearing? When such a trend occurs without a specific decision or discussion beforehand, I tend to go back to the basics. Ask the team why they think they have retrospectives: Is it just because management says they must, or is it for the sake of the team? I also ask if they have ever gotten anything out of the retrospectives or been surprised by anything they learned about each other, the cooperation

---

1. Also called "regrettospective" by Daniel Terhorst-North AKA Dan North.

among team members, or the communication dynamics that influence their work.

These questions are difficult to answer because if they already had positive answers, the team would not try to eliminate the retrospectives.

As a facilitator, you might have to remind the team of experiments and of things that happened or were shared at a retrospective. You can do this only if you have taken notes about action points, experiments, surprises, or highlights at the retrospectives you facilitate. You do not need to take copious notes during the retrospective (you will not have time for that), but you can make short notes about big things and after each retrospective write down what they decided to try.

You could also restart the retrospectives with new activities, a new facilitator, perhaps an external facilitator, as described in the **Do It Yourself** anti-pattern (Chapter 10). Sometimes it helps to get management to back up the decision of having reflection with retrospectives. Norm Kerth once very wisely said to me: "You cannot sell retrospectives, nobody wants retrospectives. But you can sell solutions to the problems they already know about." Of course, by *sell*, he meant "convince people that it is worthwhile."

An assessment, such as the Agile Fluency diagnostic (Chapter 7, **Nothing to Talk About**), could also be used in this situation to enable the team to visualize where retrospectives might be most helpful.

ONLINE ASPECT

Since this antipattern is about planning how long the retrospective should be, the difference between an online and an offline retrospective is not significant in this context. Bear in mind that because of the often short attention span of human beings, online retrospectives must be shorter than offline ones or should have a break or some break-like structure built in with time set aside for laughter. This time for breaks is often the first thing some people want to get rid of in order to save time for "real" work, and you might have to insist on keeping this extra time in the retrospectives. As in an offline retrospective, you have to convince the team that the time spent on retrospectives is valuable and that breaks are there for a reason.

PERSONAL ANECDOTE

I have often been in this **Get It Over With** antipattern. It tends to happen when the same facilitator and team work together and repeat the same activities over time. The retrospectives seem less rewarding and consequently are seen as a waste of time—or at least as a use of time less valuable than actual coding.

In one situation, I had heard a rumor that the team got very little out of the retrospectives. I decided to start the next retrospective by asking them what they expected to get out of these next 1.5 hours. From two members the answer was "nothing." Had I been a less experienced facilitator or known these people less well, this answer would have freaked me out. Instead, it kindled my curiosity and my drive to make this a very good retrospective.

I made them very aware every time they decided to discuss or not discuss something. I also made sure that they had one to three actionable experiments with them when they left the meeting, that someone was responsible for making each experiment happen, and that a time frame had been established for each experiment. After the retrospective, I asked what the team thought they would get out of these experiments, and since the ones they had decided upon could be traced back to issues and challenges, it was an easy answer. In my experience, even the most code-driven developer can be won over if the retrospective is visibly valuable. As can be seen in Chapter 20, **Negative One**, the hardest critic can turn into your best champion for retrospectives.

For another team, I had been facilitating their retrospectives for years, and yet the feeling of not accomplishing anything crept in on them. For the next retrospective, I took all the experiments they had decided upon in the last 6 months and lined them up on the whiteboard, one Post-it Note for each experiment. Then I drew a figure on the board with some space for items that were done, some space for items that were now irrelevant, and a bigger space for items that were still relevant. In the space for the relevant items, I made a vertical line with "Long term" at the top for the long-term goals and "Action points" at the bottom for experiments that could be started today. See Figure 12.1.

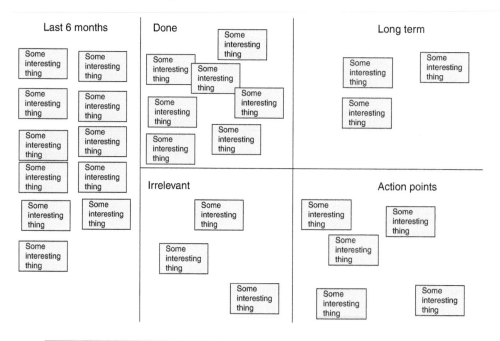

Figure 12.1 Action points from the past retrospectives

The exercise with the team was now to go through all these Post-it Notes and figure out what category they should be in. Most were categorized as Done, and the team had a short talk about what they had gotten out of those items; even small things such as "say good morning every morning" turned out to have made a difference to the team. Some of the experiments that had not been done due to timing or forgetfulness turned out to be no longer relevant. We spent very little time on those. In the end, we arrived at some experiments they could try out immediately and some that were actually too big to be carried out as they were and needed a specific meeting to figure out the best way to achieve them.

In the debriefing after this activity, we talked about how even small changes can have a huge impact and how we should be more aware in the future of the need to make the experiments immediately actionable or to plan a follow-up meeting to slice them into smaller steps.

Disregard for Preparation

. . . in which the facilitator initially misjudges how much preparation an online retrospective requires and later learns how to prepare for it wisely

Chapter 13

CONTEXT

A specialist who happens to be living on the other side of the world joins the team, and now Sarah has to facilitate distributed retrospectives with a new team member. After years of facilitating retrospectives for the team, she knows the team members, what activities are efficient for them, and what they believe works for them. It is difficult for her to see how these activities will work in an online retrospective, but she is too busy to spend time learning how distributed retrospectives are different. She sets up a distributed online retrospective by sharing an empty document and inviting the team with a video conferencing tool.

When the retrospective is about to start, Sarah logs in to the video connection and the document she asked them to use. Unfortunately, only the new specialist and two other team members are online when the retrospective is supposed to start. The other team members arrive during the next 6 minutes, but the round of questions she had decided to start the retrospective with has to be repeated twice because more people keep joining the meeting after she thought that everyone who was able to be there had joined.

When everybody finally is present in the video conference, she can only see five people in the document, and when she asks them about it, she can hear the rest desperately trying to enter the shared document. Some have misplaced the link, some have forgotten their password, and another 5 minutes are lost in this process. Having lost 11 out of 60 minutes, Sarah is a bit frustrated but tries not to let the team notice. For the Gather Data phase, she wanted everybody to add three lines about what they like, what they don't like, and what questions they have. When the third item is added to the document, the new specialist starts a long description of the issue, and Rene adds to it.

Rene and the specialist start their own little discussion, and Sarah is unable to stop it because she cannot use her body language in this online venue to direct the team's attention and indicate that they need to move on. She finds it difficult to change the setting online, since all she has is a document, which leaves her unable to throw in a new activity to make people talk in smaller groups or write more issues as she would have done in a real-life setting.

The retrospective ends when the time is up with the team having discussed only one subject. In fact, most of the people in attendance had already started looking at their email before the retrospective ended, and there was no real wrap-up with action points, experiments, or lessons learned. The momentum of the retrospective, if there ever was one, was lost.

GENERAL CONTEXT

It is becoming increasingly common that teams are geographically distributed and retrospectives must be conducted online. An online setting seems incompatible with the best practices of facilitating retrospectives, since facilitating is very much about reading body language, listening to the things that are not said out loud, looking at where and how people choose to stand and whom they choose to whisper with, and so on. All of these activities are harder or impossible online. Also, an important part of a retrospective is that people show respect for each other by not using their phone or reading email. This expected courtesy is hard to implement in an online retrospective.

In addition, an online retrospective must be kept short. I normally do not make them longer than 45 to 75 minutes. It is harder for people to keep their attention on the retrospective without the physical aspect of walking around that normally is part of a retrospective. Although I think this might be solved with future technology, we are not there yet. Also, scheduling longer online meetings is difficult: due to the distribution, people may be in locations where they have conflicting all-hands meetings, or they may be in different time zones, so the retrospectives must be arranged to accommodate people's need to eat, sleep, pick up children from kindergarten, and so on.

Disregard for Preparation is an often recurring antipattern for distributed retrospectives and has huge consequences. Sometimes, the retrospective facilitator is asked to facilitate an online retrospective without much notice, perhaps because the original facilitator is indisposed, and the new facilitator has little time for preparation. In addition, if the facilitator has never facilitated an online retrospective before, he or she might not be aware of the added challenges inherent in this setting.

ANTIPATTERN SOLUTION

What sometimes happens is that a retrospective is set up with an invitation to a video conference tool (Meet, Zoom, Skype, Teams, etc.), and a document (Google Drawings, Miro, Mural, etc.) is shared between the invited people. Perhaps someone has talked about a theme for the retrospective, but otherwise, it is often an example of "Just show up—we will figure it out."

CONSEQUENCES

The only reminder for an online retrospective is whatever technical reminders are set up in the calendar. In a real-life meeting, you might be prompted by seeing your colleagues get up and go to the meeting room. In consequence, people come at the very last minute to the online meeting or, too often, a few minutes late.

As a result, the facilitator starts with 5 minutes less than planned in a retrospective that perhaps was originally just scheduled for 60 minutes with up to 10 people. And still, the facilitator is expected to Set the Stage, Gather Data, Generate Insights, Decide What to Do, and Close the Retrospective in the time remaining. Without preparation, it might look like this: "Okay, welcome to the retrospective. Who has something they want to share?" Someone starts talking about a subject close to his or her heart. If the facilitator does not stop this person in a reasonable amount of time to hear what others want to share, the retrospective could just revolve around the subject that the first person brings up.

In a real-life retrospective, the facilitator would more easily be able to detect that others are waiting to share and make sure people are aligned in their discussion, because it is easier to read people in a face-to-face meeting and to change the activities accordingly. Of course, these retrospectives also need preparation, but the facilitator's job is simplified when he or she can observe the dynamics taking place in the room and find activities that fit the situation.

If a retrospective becomes focused early around one particular subject, this may be a very important subject that someone genuinely needs to discuss, but no one else is heard, and there is no consensus that this is the most relevant topic for them to discuss.

If the retrospective is not wrapped up on time, it can nevertheless come to an abrupt end as people suddenly log off and go to other meetings, since it is much easier, socially, to leave an online meeting than to leave a physical room. In the worst-case scenario, these online retrospectives consist of either speeches from the **Loudmouth** (Chapter 18) who likes to talk the most or a discussion between a few team members with no full agreement on experiments. They become status meetings, the soul-destroying evil twin of the retrospective.

SYMPTOMS

The retrospective starts later than planned because people are late. When they arrive, they have trouble logging into the document, or they don't share video (see the **Peek-A-Boo** antipattern [Chapter 16]), or they mute themselves. Some people stay quiet because there is no activity planned for them to share. The agenda is impossible to follow, and the retrospective ends abruptly.

REFACTORED SOLUTION

Prepare for the retrospective in a number of ways. For example, make a shared document,[1] and send an email at least 1 day before the retrospective with a link to the shared document and ask team members to make sure they can access it. If the team was warned at the last retrospective that they were expected to prepare for the next retrospective by populating the shared document with virtual Post-it Notes, you can remind them when you send the link. You can also ask them to start filling in the document before the retrospective. In that way, you can save some time in the retrospective, and it might be easier for some people to fill in the document about positive and negative events when they can sit quietly alone with their calendar. Then, when the retrospective starts, you can set aside some time for them to read all the inputs. With this approach, everybody is prepared. It works well and has been implemented by companies such as Amazon, where the first 10 minutes of every meeting are set aside for people to read the document in order to have everyone prepared and in the right mindset for the meeting contents.

Send an email again on the day of the retrospective 15 minutes before it begins, reminding the team that this is a good time to get coffee. Otherwise, people forget until the very last moment. Then they will want to get coffee, and then when they get up, they notice they need to go to the toilet. At this point, the first 5 to 7 minutes of your retrospective have already been wasted.

It is also important to prepare a detailed schedule for the retrospective. Be aware how much time you have for each phase in the retrospective, and do your best to stick to it. If you can see it is not enough time, ask the team what they would like to do about it. Extending an online retrospective is usually not an option, so the realistic options are to choose only one of

---

1. I prefer Google Drawings because it gives more freedom on the fly, when you need to change something quickly, and because it is easy to use Drawings to simulate using Post-it Notes on a board. But boards can also be made in Miro, Mural, and Trello.

the subjects on the board to discuss or to arrange a follow-up retrospective. Either option is a better alternative to just hurrying to the end of the retrospective and letting everybody log off without a wrap-up.

As always, have a backup plan for the retrospective, another agenda that allows for a change without demanding a lot from you. Retrospective facilitation, like most software development, should be agile, as described by Joseph Pelrine (2011), and plans should be followed by actions and feedback loops. This is described in more detail in the Cynefin framework (Kurtz & Snowden 2003), where *complex systems* should be dealt with in a probe-sense-respond fashion: you try something out, sense what happens, and respond to the reality instead of following a complicated plan.

Preferably, everyone should have his or her own camera and be in separate rooms. This can be hard to achieve, but it prevents in-real-life subgroups of people from having a parallel discussion on their own.

If you choose to have a physical board in one of the rooms, instead of a shared document online, you might want to give each person an *avatar*, also called a *proxy*, someone who acts on behalf of another person who is not physically present. Set up a phone call or a chat between two people and let the avatar write Post-it Notes for the person he or she represents.

In summary, you should do your best to make everybody equal at an online retrospective, even if you feel the odds are initially against it. One benefit of online retrospectives is that it seems more natural to make all discussions using the *round robin* technique, instead having free-flow discussions in the plenary.

Round robin is a pedagogical pattern, described by Bergin and Eckstein (2012), that I use extensively in my teaching.

What Is a Round Robin?

A round robin is an approach to asking people to speak in turn to ensure that every voice is heard. Everyone gets to answer a question, choose a subject, or comment on something, one after another. If they are sitting around a table, it could follow the seating arrangement; if online, it could follow the list of names you have for the people in attendance. Sometimes, it is necessary to set a time limit on each person's turn—for example, allow each person one minute—but most of the time, people are mindful that time is limited and that everyone has to say something in the round robin. It is sometimes difficult to make people wait for others to stop speaking because they are afraid that they might never get to share their input. With a round robin, everyone knows that they will get a chance to speak, so they let others speak without being worried that they won't have a turn.

There is a hidden danger in the round robin: sometimes, one person says something that you think is really good, and in that case, you might feel tempted to thank the person or point out how great his or her input was. Try to avoid that temptation, though, and make everything said equally important or interesting to avoid discouraging anyone from contributing. You can also choose to thank everybody after they have said something. But choose all or nothing.

ONLINE ASPECT

Since this antipattern is in the context of an online retrospective, everything applies to an online retrospective. For offline retrospectives, much of the advice in the refactored solution can be helpful. For example, it is still a good idea to remind people about the retrospective a day in advance, especially if there are some experiments the team decided to try out. Then the person responsible for giving feedback on how the experiment ran has some time to prepare. Making everybody equal is also important for an offline retrospective, and this can be done by making sure that everybody is heard, as described in Chapter 18, **Loudmouth**, and Chapter 19, **Silent One**.

PERSONAL ANECDOTE

I was invited to facilitate an online retrospective. I spent some time talking to and emailing with the person who invited me in order to make the best retrospective possible given the circumstances. In most cases, when I facilitate an online retrospective, I do it from my home office, but in this case, I was asked to facilitate from the company's premises.

When I arrived for the distributed retrospective, three people were in the same room as me, two others were together in another location, and the last two were alone, so we were eight people in four locations. One of them took the meeting from his phone, and, as we discovered later, he was in a coffee shop. Had this been a meeting where the team needed only a one-way flow of information from me, their scrum master, or from their manager, it would probably have been okay. But in a retrospective setting, no one person is more important than any other, and everybody should have an equal chance of both listening and being heard.

The usual thing happened: the people I could not see had also chosen not to see us, so we were only audio to them. The coffee shop participant was mostly invisible and fortunately muted due to coffee bean grinding, and the other was in front of her computer. As a result, when I asked for feedback or stories behind something, I always had to ask them twice. The typical "Oh, please repeat that, as I didn't get it" or "Ah, you were talking to me; please say it again" were heard.

This is very typical for unprepared distributed retrospectives: you may have set up a document and a virtual meeting, but you have not prepared the people attending for what they need to do in order for everyone to gain from it.

When I am an ordinary participant in a boring online meeting myself, I sometimes play Solitaire when I am just listening. I have learned that it takes the edge off the boredom while still keeping me alert enough to jump in when needed. (I would never do that as the facilitator, naturally.) My husband watches 30-minute-long fractal zoom videos on YouTube during boring meetings. If you try this, remember to mute the sound!

Suffocating

. . . in which team members get tired and hungry and unfocused during the retrospective, and the facilitator makes sure to feed them and give them oxygen so that they can concentrate a bit more

Chapter 14

CONTEXT

It was one of those winter days in Denmark when you couldn't wait to get inside. It was only just possible to get to work,[1] and there was a veritable snowstorm outside. The old Danish saying, "There is no bad weather, only wrong clothing,"[2] still held, if you were willing to dress up like a polar explorer.

In the retrospective, everybody enjoyed looking at the snow outside the windows, but opening the windows would not have been a popular idea. So, the retrospective went on with closed windows, and this was one of the very long retrospectives, covering a whole project. After about an hour and a half, people in the room were becoming drowsy; they seemed unengaged, and they were a bit irritable toward each other. After 2 hours, Peter and Andrea were asleep, and the rest of the team wanted to stop the retrospective. Sarah decided to stop the retrospective, and they agreed to continue another day. After lunch the storm stopped, the sun came out, and they spent an hour outside, making snowmen and snow angels (Figure 14.1). And they completely forgot to continue the retrospective another day.

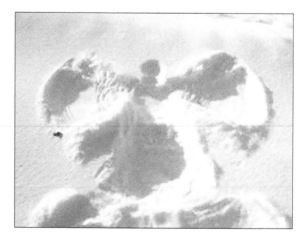

Figure 14.1 A snow angel

---

1. The most stubborn people were still able to cycle to work. You can Google pictures of people cycling in the snow, if you doubt me.
2. It rhymes in Danish: *Der findes ikke dårligt vejr, det hedder kun forkerte klæ'r.*

GENERAL CONTEXT

We sometimes forget that we are animals. Working with our brain all day, we tend to ignore our bodily needs. We need oxygen, food, and restroom breaks; women generally need to be in a warmer room than men; and let us not forget about smokers, who need a cigarette now and then in order to be able to think about anything else. When we forget all this, we sometimes try to push ourselves to come up with creative ideas or discuss painful events despite being in a bodily state that does not allow us to do so optimally.

ANTIPATTERN SOLUTION

The antipattern solution is to carry on with the retrospective in order to just get it finished instead of calling for a break. It often seems obvious after the fact that you should have opened a window, brought snacks, or taken a short break. The trouble is that you need to plan for this before you yourself get hungry, dehydrated, oxygen-deprived, or in desperate need of a toilet, because once you are affected, you might not be able to help other people.

CONSEQUENCES

Without oxygen in the room, people can become sleepy and even get a headache. Also, if the retrospective is held during a part of the day when you would normally have lunch or an afternoon snack, people might become hangry,[3] uninspired, or unable to concentrate. The consequences are that the team members are unable to listen to each other or follow the Prime Directive while communicating, and the retrospective will be a waste of time.

SYMPTOMS

If you see a participant fall asleep or someone getting really irritated over seemingly small issues, this could be a sign that the human needs are not met.

---

3. Anger stemming from hunger.

REFACTORED SOLUTION

The refactored solution is simple, for once. Open a window, if the weather permits and the rules allow (I am told that you can never open windows in US office buildings). If it is not possible, then either make sure that the room has been ventilated before you start the retrospective or invite people outside after 45 minutes. Make sure not to plan the retrospective during lunch, and if you can't avoid a lunchtime meeting, then bring some healthy snacks. Unhealthy snacks can also be useful, especially in the afternoon or if you expect painful[4] discussions. Some organizations or teams have decided on working agreements to establish, for instance, that a break must be included in every meeting longer than 45 minutes.

One way to make sure that there are breaks when needed is to make everyone either stand up or sit down. If you are the only one standing up, while the team is sitting down, you won't notice as quickly that the team's energy level is low.

It is important that you, as a facilitator, plan breaks in advance, because you might be in a situation where you are not able to think about it. For every retrospective, then, consider oxygen level, hunger, pain, and breaks. The Do Food pattern from Linda Rising and Mary Lynn Manns's book *Fearless Change* (2005) supports these considerations.

ONLINE ASPECT

If the retrospective is online, you cannot bring food for attendees or open their windows, so you have to plan your way around this limitation. Be mindful about how long people can be expected to focus on an online meeting—probably not more than 45 minutes without a break. But a break can be many things: it can be laughing together, for example, or standing up and doing something physical. It does not have to be a break whereby everyone leaves the meeting and comes back. If you want the team to attend a retrospective late in the day, you can encourage them to

---

4. Sugar is a natural painkiller.

bring a snack. A presentation of the snack can be a fun icebreaker for the start of the retrospective. If you do this more than once, the team members might even become creative with their snacks to get a few laughs or some envious looks at their Ziploc bags.

PERSONAL ANECDOTE

I have to admit, this is one of the antipatterns I find myself in from time to time. I am always so focused on the planned activities, the energy in the room, and communication (verbal or nonverbal) between the people in the team that I tend to forget that we have more primitive human needs.

Add to this that I am very sensitive when it comes to blood sugar levels. If my blood sugar drops, I become unable to think and then hangry. Often, I notice a change in behavior in myself before I notice it in others. I might silently start hoping that people will stop talking, because what they are saying is utterly stupid and irritating. Or I might notice that I want to add ironic or sarcastic comments to everything people say. Now, irony and sarcasm are not necessarily bad, but if you use them too much, they start to sound like criticism.

In my dream scenario, I always book the retrospective room 15 minutes before the retrospective so that I have time to open the windows and let some oxygen in and to prepare anything else I think we will need. This could be arranging the chairs in a circle, putting pens and Post-it Notes on the table, or writing the Prime Directive on a poster.

I try to remember to ask the sponsor (my word for the person who invited me to facilitate a retrospective) to bring snacks for late-afternoon retrospectives. When the sponsor puts them on the table, he or she often gets a lot of credit for such thoughtfulness. In some situations, it can be important to allow the sponsor to keep the credit, even if you think you earned it, if you know he or she needs it more than you do. The main point is that the team members are happier and more able to focus, learn, and communicate in a pleasant way.

Curious Manager

. . . in which a manager is curious about what happens at the retrospectives and wants to listen in on them, and the facilitator, in a nice but firm way, says no to the manager

Chapter 15

CONTEXT

The team's new boss, Janni, is curious about what is going on at the retrospectives. She has noticed that there is always a change after a retrospective, and she is often asked to do something for the team. She believes that if she could be present at the retrospectives, she would be able to help them more quickly and effectively. Sarah understands Janni's reasoning and allows her to attend the next retrospective.

But something weird happens. Andrea and Peter are quieter than usual. When she asks them why, their answers are vague. The retrospective is going through its stages as it should, and everything seems to be fine—except that only very superfluous problems are discussed, and the whole room is a bit tense.

This antipattern is described as *Line Managers Want to Attend* in Luís Gonçalves's blog post on antipatterns for retrospectives (Gonçalves 2019).

GENERAL CONTEXT

It is often the case that the boss or manager becomes curious about what is going on behind closed doors. I have sometimes heard managers say, "I need to be there to hear what really happened," or even, "I'd better be there and set this straight—they really need to stop making mistakes and to work faster." Sometimes, this is perfectly fine, and there is trust between the boss and the people on the team, but most times, in my experience, the team feels more at ease without the boss in the room.

Building trust in a retrospective setting can be difficult, but destroying it is easy. It is one thing to invite the manager to the retrospective because the team wants her to be there, as in one of the refactored solutions to **In the Soup** (Chapter 3). But it is quite another if the boss wishes to be there, and the team is not asked for permission.

ANTIPATTERN SOLUTION

Perhaps the manager is allowed in on a retrospective for the team, just to observe. The manager may even promise not to say anything or disturb the retrospective—she would just really like to "see the magic." The facilitator, either understanding the boss's view or being afraid of her, accepts and informs the team that the boss will be present at the next retrospective.

CONSEQUENCES

Several different consequences can result from this antipattern. One could be that since the team members all feel safe around the boss, they behave as usual, and the boss gets a rare view into the dynamics, fears, and hopes of the team. More often, however, the team becomes anxious when the boss is present because she is responsible for hiring and firing, promotion, and pay raises. Team members fear that if they reveal their concerns and problems that have occurred, they might lose their job. Also, the boss might not be able to keep as invisible as promised, and once she starts talking, it can be difficult to get her to shut up again!

SYMPTOMS

Some person outside the team invites him- or herself to the retrospective, disregarding how the team feels about it. Some team members are quieter than usual. The team only talks about the positive things.

REFACTORED SOLUTION

One simple solution is to keep bosses out of the retrospectives. Actually, allow only the team at the retrospectives. If bosses or managers are curious, throw a separate retrospective at the management level. My rule of thumb is this: "If you can hire or fire, stay out of the team's retrospective." Now, of course, you have to explain to bosses why they are kept out of the meeting. Some bosses may fear that the team wants to talk about them behind their backs. You can reassure them that a good facilitator always tries to discourage venting against specific people and instead directs discussions to focus on communication or cooperation. Also, if managers fear what might be said about them, then this might be a symptom of another problem. Maybe the organization needs to work with trust issues, or maybe the managers need coaching.

During the retrospective, you can ask the team if you are allowed to share some of the information that was visualized at the retrospective, such as Post-it Notes or posters. Doing so can have the added benefit that the team becomes aware of what they would like the manager to know that they might not be comfortable relaying directly.

Alternatively, you could ask the team for permission to invite the boss, because sometimes they are okay with it. Asking can be tricky if people are afraid to say they would prefer not to have the boss attend the retrospective. Perhaps an anonymous vote is appropriate here, but be aware that every need for an anonymous vote is actually a sign that there is not enough trust among the team members.

ONLINE ASPECT

For an online retrospective, all the same challenges and solutions apply. One difference is that an online retrospective often can be recorded, and the online documents can be shared easily. It is important to stress that everything that goes on in a retrospective stays in the retrospective, and anything that is shared is shared only if every participant is agreeable to doing so.

PERSONAL ANECDOTE

I once worked in a big company as an agile coach, and I had several teams for which I facilitated retrospectives. I had managed to keep the managers out of the retrospectives, but it took some convincing for them to understand. One of the teams was too afraid to say no directly to the manager, and I had to say it was against my principles. The team told me they did not want her there under any circumstances, but when she asked them, they were all smiles and "of course," so I had to be the bad guy with my "stupid" principles. But that is the role of the facilitator at times, and I have learned to live with it.

This went well for a while, but this was a team with huge problems, and one of the team's problems was this very manager! She frightened them, but I had to keep my promise to the team that, as in Las Vegas, everything that is said in this retrospective stays in this retrospective. Consequently, I could not even let the manager's superior know about the problem. At one point, the manager dragged me into a room and asked me to spy on the team for her in a retrospective. She said, "You are so good at making people talk, so you need to make them talk about this and let me know what they are saying." I have to admit that I also was afraid of her, but I had to tell her that that was simply not possible and that if I told her what the team said, I would quickly lose my ability to "make people talk." She understood my position but continued to ask questions after each retrospective and even asked some of the team members as well. I later learned that she left the company.

Peek-A-Boo

. . . in which team members will not show their faces on the video in an online retrospective, and the facilitator learns why and finds ways to make it safer for people to show their faces

Chapter 16

CONTEXT

Sarah has prepared for the next online retrospective with a shared document, and she has reminded all of the participants that there is a retrospective and which document they will be using. Everybody but Bo is online in the meeting on time. It turns out that Bo used the link from the last retrospective instead of the one in the calendar invitation, but this is soon sorted out via chat, and everybody is present 3 minutes after the scheduled start time. Sarah had asked the team members to turn on their video in the invitation to the meeting. She quickly realizes that only half of them are on video, so when Rene, Andrea, or Kim is talking, only a black screen is visible to the participants.

Sarah starts the retrospective with the usual round of questions for everyone. Then she starts to gather data on the shared document with virtual Post-it Notes. While she can see most of the team on video and thus evaluate whether they are concentrating on the task, she is unable to judge whether Rene, Andrea, and Kim are writing or doing something else. After 10 minutes, she asks the team to look at what was added to the document and to say out loud what they think. At first there is silence. Everybody is waiting for someone else to start speaking, and then, when Andrea begins, everybody else starts speaking at the same time.

With some difficulty, Sarah facilitates the rest of the retrospective into an agreement on a new experiment for the team to try out until the next retrospective, but she has a sneaking feeling that Kim never really participated and perhaps was occupied the whole time with a task unrelated to the retrospective.

GENERAL CONTEXT

In online retrospectives, many people choose not to appear on video. There can be many reasons for this choice, and we will dive into them later. If you are like me, you find it much easier to facilitate a retrospective when you can see the people who are participating. Their facial expressions give you small clues as to when to change the activity, when to slow down, or when to speed up the discussion. Their expressions can also warn you right before they become angry or choose to opt out of the retrospective. However, when you have no input other than an often muted microphone, it is really hard to do your work well.

ANTIPATTERN SOLUTION

A facilitator often allows the participants to opt out of the video. This could be because the facilitator wants to be nice and not force team members to do something they feel uncomfortable about. Or because the facilitator fears not getting to facilitate the retrospective if he or she makes demands. Or simply because the facilitator does not see the need for it him- or herself.

CONSEQUENCES

Without being able to read facial expressions, it is hard for the facilitator and everyone else at the retrospective to see when someone is bored, getting angry, sad, or simply wants to say something. Even in an offline retrospective, silence can be hard to interpret, but someone who is silent and also not visible is extremely difficult to read. One of the consequences is that it is easier for the participants to hide and turn their attention to other tasks rather than take part in the discussions during the retrospective. By doing so, they cannot take part in decisions, and the team can end up deciding on actions that the nonparticipants think is a bad solution. In addition, their colleagues do not have the benefit of the nonparticipants' input, and the sharing of their different experiences to create a whole picture is lost.

The most aggravating consequence for the facilitator is that the nonparticipants often do not listen, so questions or whole parts of discussions need to be repeated for them. This unnecessary repetitiveness can destroy the planned agenda, which is an important part of a retrospective, particularly an online retrospective. More important, team members who do not participate show disrespect to their colleagues by not listening while their teammates share thoughts and experiences.

SYMPTOMS

The most obvious symptom is that people have not turned on their video, so they appear as a black screen, perhaps with their name or initials on it. Because people find it easier not to be mentally present at the meeting, you will hear comments such as "Please repeat that" and "Oh, were you asking me?"

REFACTORED SOLUTION

As in other antipatterns, the right solution depends on the context. The context, in this case, is the reason the participants choose not to appear on video. You could, of course, ask them during the retrospective, but you might not get the real answer when they have to answer in front of everybody else. As with the antipatterns that revolve around personalities, you have to tread carefully here and perhaps ask them directly outside of the retrospective or make it possible for them to answer anonymously. There is also, as always, the possibility that they do not know why themselves—it just does not feel right for them.

The first thing to do is to explain to the participants why it is important that they share a video of themselves with you and the other people in the retrospective. Reasoning with them might convince some to start their video, because they perhaps had not considered that people would want to look at them. I presume that my readers are all very beautiful people and want to be seen on video, but not everyone sees themselves in a positive light.

I often hear people say that they don't want to be on video because they are in a coffee shop or in a car, and it is not possible. This excuse for not being on video is one I do not accept. It shows a much bigger problem that has nothing to do with video. A retrospective should be taken seriously and given 100 percent of the team members' attention. It demands focus in order to be valuable. This is also why I ask people not to be on their phones or computers while we have an offline retrospective.

Some people say that they do not want to share their surroundings with their colleagues. In these situations, I try to make the team use a collaboration tool that offers the possibility of a blurred background, so that only the person is visible. There are many reasons for not wanting to share a home office with colleagues. Perhaps it is messy, cheap, or overly luxurious; perhaps political statements or risqué art decorate the wall, a number of weird books line the shelf in the background, or a spouse is walking around half-naked. No matter what it is, it can be blurred out.

Then there are those who choose to work from home without makeup or without shaving or without clothes on. This could be a part of another problem. Some people find it hard to motivate themselves to work from home, and in these cases, it can be helpful for them to dress up as they would do for work to get into work mode. For some, it is enough to just dress.

Some people do not like to see themselves on video, and since most collaboration tools show you yourself on video together with everyone else, that can feel uncomfortable for them. The reason we see ourselves on video in such a tool is that we want to make sure that we are not doing any of the awkward things we might do when we are alone. Seeing ourselves helps to keep us civilized. Those who do not fancy looking at themselves during the retrospective can opt to hide the self-view. If not, they can change the tool they use or can use a Post-it Note to cover that portion of the screen.

I acknowledge that 6 hours of video meetings each day can be exhausting—I have experienced that myself. I sometimes change my video meetings to phone meetings, especially if it is only with one other person, and if possible, I take a walk outside while we talk. For retrospectives, though, I always use video due to the importance of body language.

I recently learned of a theory for why it is exhausting to have long video meetings. It is based on the fact that *cognitive dissonance* makes us uncomfortable. When we watch people talk with us but are not actually together with them, our brain sends contradictory signals, and we feel uncomfortable in our attempt to make sense of the situation. The two worldviews—that we are together and that we are apart at the same time—create a contradiction in our mind.

Cognitive Dissonance

Cognitive dissonance occurs when a person holds two or more contradictory beliefs, ideas, or values or participates in an action that goes against one of these, and experiences psychological stress because of that contradiction. According to this theory, when two actions or ideas are not psychologically consistent with each other, people do all in their power to change one of them until they become consistent. The discomfort is triggered by the person's belief clashing with new information perceived, wherein they try to find a way to resolve the contradiction to reduce their discomfort.

Another interesting example of cognitive dissonance is that if you do a favor for someone whom you dislike, you experience uncomfortable dissonance until you start liking that person better. The logic of your brain is that if you do that person a favor, you must either like the person or be crazy. And the brain chooses the former option most of the time. This is also known as the *Ben Franklin effect*.

Ben Franklin Effect

"Having heard that he had in his library a certain very scarce and curious book, I wrote a note to him, expressing my desire of perusing that book, and requesting he would do me the favour of lending it to me for a few days. He sent it immediately, and I returned it in about a week with another note, expressing strongly my sense of the favour. When we next met in the House, he spoke to me (which he had never done before), and with great civility; and he ever after manifested a readiness to serve me on all occasions, so that we became great friends, and our friendship continued to his death."

—Benjamin Franklin

In summary, find out what problem interferes with each individual who is reluctant to appear on video and try to solve it. If you are unable to convince people to turn on their video or you learn that there is a good reason for them not to be on video, ask them to at least upload a picture of themselves into the collaboration tool so that you can talk to their face and not their initials.

Of course, there are exceptions. If people rush to the meeting and it is either no video or no attendance, the attendance without the video is preferred.

ONLINE ASPECT

Since this antipattern is found only in the context of online retrospectives, everything described in this chapter applies to these.

PERSONAL ANECDOTE

I had been facilitating retrospectives for a team for more than a year. The team was co-located all the time, and all retrospectives were done with everyone in the same room. Now times changed, and they all had to work from home.

We started doing our standup meetings online, and I noticed that some of the team members chose not to show their video. For the retrospective, I asked them to be on video, and most were on video in the beginning, but when they noticed that not all were, they disappeared one by one.

I asked them why they chose not to be on video, and they gave me multiple answers. For one, the bandwidth from home was insufficient. I wondered about that, since it had already worked for some meetings, but I chose not to question it. Another person did not want his colleagues to know anything about his private life. For him, these were two very different worlds that should never meet. I had always sensed that this person shared a lot less about himself with the team than did the others, and now I learned why. I believe that we go to work as whole human beings and that who we are at work and who we are in the rest of our lives greatly influence each other. It was interesting for me to learn that some people are unconvinced that such an influence exists. A third said that he disliked having people's faces so close to him on a big screen, that it made him uncomfortable. I thought about mentioning that he could minimize the window or put it underneath another window, but I felt the time was not right.

Since I assumed this situation was not permanent, I decided to let it go and just asked those reluctant to appear on video to upload a picture of themselves in the tool so that at least I could look at that when I talked to them and when they talked to me. If this was a team where I knew we would be working online permanently, I would have tried to be more persuasive. We had three more retrospectives like this, and most of them were on video most of the time. Sometimes you just have to choose your battles with a team, and I had enough to work with here.

I have had numerous retrospectives where everyone is on video, and I have had a lot of interesting discussions and a lot of fun. Most of the teams I have worked with have come to the understanding that it helps the team when everybody is visible.

Sometimes, a video meeting can even have benefits that a meeting in real life does not have. Figure 16.1 is an example of me facilitating a Six Thinking Hats retrospective. The screenshots are from the time when a Google Hangout could enable you to put on different headgear. I also used a beard sometimes to hide my double chin. I really miss that.

Figure 16.1 Here I am facilitating a Six Thinking Hats retrospective with different hats.

People Antipatterns

Disillusioned Facilitator …in which the team mocks the facilitator for using ridiculous activities, and the facilitator explains why the activities are useful

Loudmouth …in which a team member needs to hear him- or herself all the time, at everyone else's expense, and the facilitator applies various tactics to ensure the rest of the team is heard

Silent One …in which a team member chooses to be almost completely quiet, and the facilitator applies various tactics to make sure the Silent One is heard

Negative One …in which one team member's attitude has great negative impact on a retrospective, and the facilitator shields the other team members from the negativity

Negative Team …in which the team wants to talk only about the negative things because they think these are the only things they can learn from, and the facilitator shows them that a focus on positive aspects can be equally valuable

Part III

Lack of Trust ...in which the team members do not trust each other enough to share anything of importance in the retrospective, and the facilitator helps them build that trust

Different Cultures ...in which the assumptions the facilitator or the team members bring from their own culture are preventing them from seeing how the retrospective is experienced by others, and the facilitator finds ways to make them more aligned

Dead Silence ...in which the team members are completely silent, often in an online retrospective, and the facilitator uses various tactics to hear their opinions despite their reluctance to participate

Disillusioned
Facilitator

. . . in which the team mocks the facilitator for using ridiculous activities, and the facilitator explains why the activities are useful

Chapter 17

CONTEXT

Sarah is still inexperienced as a facilitator, and at the next retrospective, she has decided to introduce a new activity: every retrospective should start with a round in which, prompted by a question, everybody shares something from their personal life. The question could be anything, but she decides to make it simple for the team and start with, What did your last meal consist of?[1] Sarah believes this is a safe question that doesn't require much thought and that people will be comfortable answering it.

Although she feels this activity will be useful, Sarah is a bit worried about asking people to share something personal, because she has overheard some team members say on occasion that they are not interested in each other as people, only as colleagues. When she starts the retrospective, she asks everyone to stand in a circle and describe their last meal. People start looking around in a puzzled way, someone giggles, and most of them look at Rene.

Sarah has noticed before that even though Rene is not the leader or the manager, he seems to be the "natural leader" of the team. Whatever he does, the rest of the team wants to do as well. And it is obvious that he does not like this activity at all. In earlier retrospectives, he has called the activities "games" to emphasize how childish and unimportant he finds them. (Actually, in my opinion, playing games and being childish are important, and there are good reasons[2] for doing so.) Rene says that it is a silly thing to waste time on now, when they are so busy with the implementation of the new API, so perhaps "we should call it a meeting and go back to the real work." Everybody leaves—everybody but Sarah, who stays, disillusioned.

---

1. Don't ask this question in Denmark, though. It is boring because everybody eats cheese sandwiches and oats for breakfast. It is more fun in a distributed retrospective with people of different nationalities.
2. You can read more about this on Portia Tung's The School of Play website (2019).

GENERAL CONTEXT

Many facilitators, especially at the beginning of their journey as facilitators, find themselves initiating an activity despite being worried or uncertain about the activity and unconvinced that it will work. They might disregard their concerns because someone recommended the activity or because it was described in a book or a blog post about effective retrospectives. Perhaps it was even described in the book *Retrospectives Antipatterns*, in which case it should be a great activity.

ANTIPATTERN SOLUTION

The antipattern solution is to go ahead with the activity, but somewhat reluctantly and almost apologetically, saying, for example, "I know this sounds stupid, but people say this activity is good" or "I am not sure about this activity, but the book says we should do it like this."

CONSEQUENCES

The consequence can be that no one else takes the activity seriously either, because if the facilitator does not seem to think it is a good idea, it could just be a waste of time. The team might try it out halfheartedly—all but guaranteeing its failure. They might make fun of the idea or even make fun of the facilitator. The long-term consequence is lack of respect for the facilitator and for retrospectives, and in that case, you might as well stop doing retrospectives for this team. This means that the team will not be learning what they could from the events they experience in their daily work. Last but definitely not least, the facilitator will lose even more self-confidence.

SYMPTOMS

The symptoms can be found in yourself: you feel uneasy about facilitating a particular activity, one that you find yourself not really believing in. You might also notice symptoms among the team members; for example, they start giggling and whispering to each other or refuse to do what you ask them to do. Another symptom is that proposed activities are abandoned without being properly attempted.

REFACTORED SOLUTION

"Fake it 'til you make it" is the short answer, but there is more to making an activity successful than just putting on your best facilitator's smile. Sometimes you will come across an activity that you are not sure will work in this particular setting with these particular people. And perhaps you are right, but it is certain that if you do not believe it will work, then it will not work.

People usually will do what it takes to make the retrospective a good experience. They might not want to admit it at first, but even though they appear to be too cool or too grown up or too serious to do something as silly as compliment their teammates, vote with their feet (that is, walk to a specific part of the room to indicate what they want), stand in a circle, or share what they ate at their last meal, they are willing to give it a try if you can convince them that it is a good idea.

And herein lies the challenge. If you are inexperienced or simply new to the team you are facilitating, you might not have earned their respect yet, and perhaps more important, retrospectives might not have earned their respect. When you decide to include an activity, make sure it is one you believe in. In the beginning, you might want to use only safe/boring activities, but with time, you will grow bolder. But choose the ones that you see and can explain the purpose of and that you do not think are silly. If you do not believe in an activity, nobody else will.

Start the activity by explaining its purpose and what you expect the team to get out of it. Then do the activity and, afterwards, make sure to point out the benefits, such as the opportunity to share, vent, identify causes, just have fun together, or get to know each other better. It's good practice to debrief the team after the activity to reinforce what they got out of it.

ONLINE ASPECT

For an online retrospective, this antipattern is both harder and easier. It can be harder to convince participants to do something that takes them out of their comfort zone because of the added border between you and them, coming from the online aspect of the retrospective. Conversely, that same border can make it feel less brutal if the team ridicules the activity. For both online and offline retrospectives, it is important that you choose only activities you believe in yourself, that you explain their purpose, and that you try to not take the negative reactions personally.

PERSONAL ANECDOTE

I once facilitated a new team, and I could feel from the start that being in this retrospective was something they had been forced to do. It was not their choice to sit down and talk about anything. They did not acknowledge anything good that any of them had done: the Gather Data phase was all about what had gone wrong, even though I tried to make them think about good events or things that gave them energy. See also Chapter 21, **Negative Team**.

I decided to apply Norm Kerth's Offer Appreciations exercise. I made the team stand in a circle, which took some convincing, but I said that it was necessary. I then explained the activity. There was a ball, and the ball would start with me, and then I would acknowledge one of them for something good and then throw the ball to that person. This person should then acknowledge something good about another person and throw the ball to him or her.

I started by acknowledging one of them for a positive comment that he had made earlier, and then I threw the ball to him. He caught the ball, looked at the ball, looked around, and slowly, without a word, let the ball fall to the ground. As I watched the ball roll along the carpet, I almost panicked. I initially did not know what to do but decided to end the activity and move on to closing the retrospective. If the same thing were to happen to me now, I would react differently.

The way I interpreted it back then was that he thought it was a stupid game, and that might still be part of my thinking now, but I also think that he simply did not have anything nice to say about anyone on the team—or if he did, he did not want to share it. That is a big red flag, which you can take as a huge gift if you experience it, because it gives you a lot of information about where the challenges in this team lie.

Today, I would probably dive into that action and ask him if he had any thoughts about acknowledgments that he discarded or if he could think of a reason for acknowledging his colleagues and fellow team members. Depending on the context, I might need to confront him with these questions outside of the retrospective instead of in front of everyone else. Making people lose face is rarely a wise choice, and if your role as a facilitator is not yet established or respected, it could also be very harmful to your relationship with the team. Another thing I would do today in that situation is ask whether someone else had anything he or she wanted to say.

At the time, I took it as a harsh criticism of my facilitation, which, in part, it was. If it happened now, I could also choose to see it as an interesting way of starting a conversation because my experience since then has given me the confidence to know that this is not (often) about me. That realization takes time.

But most important, I would have told the team why I wanted them to take part in this activity: that acknowledging one another in a team is very important because that is part of building trust, and when trust exists, people are able to speak up and ask questions. Asking questions means the problems are dealt with earlier, and the team works more efficiently, effectively, and sustainably.

Loudmouth

. . . in which a team member needs to hear him- or herself all the time, at everyone else's expense, and the facilitator applies various tactics to ensure the rest of the team is heard

Chapter 18

CONTEXT

In the development team, Rene likes to talk. He talks a lot on a daily basis, and the other team members try to start conversations with him only in places where they can get away easily, such as at the coffee machine. He loves to hear his own voice, and in the last retrospective, he spent a lot of time explaining, in great detail, the issues he was interested in.

Sarah decides to count the minutes he spends talking to compare it with how much time other people get to speak at the retrospective. Even though she expected Rene to take up a lot of time, she was surprised to find that he spoke as much as everyone else combined.

She starts noticing other team members getting even quieter at retrospectives, and some of them start focusing on their phones once Rene starts on one of his monologues. Sarah would like to call Rene out on his monopolizing the discussions, but she doesn't want to embarrass or anger him. And Rene keeps talking.

GENERAL CONTEXT

Many teams include a **Loudmouth**, someone like Rene, who seems to enjoy hearing him- or herself talk. While most of the people are mindful that a retrospective is meant to give all team members the chance to voice their thoughts and concerns, this person often uses the retrospective as an opportunity to talk . . . and talk . . . and talk. He or she might tell long stories or interrupt other people if they pause for a breath. Calling out this behavior as it happens can feel impolite or uncomfortable to the facilitator, so the **Loudmouth** is often allowed to be just that.

ANTIPATTERN SOLUTION

Let the **Loudmouth** talk, because you do not want him or her to be angry with you. This antipattern solution is used too often, perhaps because the facilitator does not know how to explain why one person dominating the discussion is a problem. Perhaps the facilitator has difficulty interrupting the **Loudmouth** because he or she does not want to be rude (an inhibition the **Loudmouth** unfortunately does not share).

If the **Loudmouth** is the only team member engaging in the retrospective, the facilitator may find it easier to let the **Loudmouth** speak than to change the activities or the focus of the retrospective.

Sometimes, people will giggle in the corner as the **Loudmouth** drones on, oblivious to the team's growing restlessness or apathy toward the tedious monologue (this is most likely to occur if the **Loudmouth** has low status on the team). More often, people just zone out when the **Loudmouth** starts talking.

CONSEQUENCES

People shut up because the **Loudmouth** monopolizes all the speaking time. They also stop listening, so if the **Loudmouth** did say anything relevant, the team would miss it.

There are two types of **Loudmouth**: the *Storyteller* and the *Breaker*.

The Storyteller is the one described in the Context section. Storytellers can't stop talking once they have started. When they have the opportunity to tell a story, they do. It might be a very relevant and entertaining story, but time often does not permit long stories in a retrospective. Just divide the number of minutes with the number of participants, and when you do, do not forget the time needed to explain activities, debrief from activities, and do other tasks, such as write Post-it Notes. This bit of math will show you how little time there is for each person to talk.

The consequence is that some people stop listening and zone out; they might start thinking about other things or even focus on their social media or emails. When the team stops listening, the result is that the common thinking you want to achieve with a structured meeting, such as a retrospective, is obstructed, and when people do not share the same focus, the meeting, whatever type it is, is not effective.

Breakers are a different species. They speak out of turn, and they have something they believe is relevant to say regarding almost every issue. No matter what other participants have experienced or felt, the Breaker has always tried something similar or better or worse. Whatever opinion another participant expresses reminds the Breaker of his or her own opinion, which must immediately be shared.

The consequence is that people are interrupted, and whatever they were sharing is lost. As a result, other team members might stop contributing altogether, because it is discouraging never to be able to finish what you want to say. This is frustrating for the other participants, and it ruins some of the atmosphere in the retrospective. It also negatively affects the outcome of the retrospective. When some data, or some insights behind the data, are missing from the larger picture, you might miss out on understanding or possible ways of overcoming challenges.

SYMPTOMS

You see people pulling themselves out of the retrospective and starting to talk among themselves. Obviously, the biggest indicator is that someone talks all the time—or at least *tries* to talk all the time.

REFACTORED SOLUTION

You could try to resolve this situation first with one of the more obvious solutions, such as introducing a talking stick that is passed around to the one who is speaking. This works for the Breaker more than for the Storyteller—it can even have the opposite effect on the Storyteller, because he or she might just hold onto the talking stick and never pass it along to the next person.

For the Storyteller, it is more helpful to allot a certain amount of time to each participant. You could use a timer or a stopwatch to measure, say, 1, 2, or 3 minutes, and then make a point of explaining how many different things there are to talk about and how many minutes are allotted for each.

Make use of variations on this magical sentence: "This is a very interesting discussion, but we have to revisit it some other time." You could also have a parking lot, a whiteboard where you save notes about what needs to be discussed outside the retrospective.

Depending on the personality of the **Loudmouth** and your relationship with him or her, there are several ways to handle his or her long-windedness. One is to plan the retrospective so that the activities include more writing than speaking. Another plan could be to keep the plenum discussion to a minimum; that way, the **Loudmouth** will always be in a small group of two or three people and thus can "contaminate" only a few people instead of the entire team.

A third possibility is to talk to the **Loudmouth** outside of the retrospective, either before, if you know you are dealing with a **Loudmouth**, or after, if you were unprepared for it. The most important advice when dealing with personality issues is that they are usually best dealt with one-to-one and not in plenum. If you ask people in front of a group to change their behavior, you are likely to get a bad result. Most people feel threatened, embarrassed, or exposed if you confront them with their unwanted behavior in public, and they are likely to react with anger, frustration, negativity, or sadness. By contrast, a private conversation shows respect for the **Loudmouth**, and he or she is more receptive to changing the behavior.

Sometimes, explaining the effect their behavior has on other people is an eye-opener for **Loudmouths**, since they might not even be aware that they are talking so much. If you think that may be the case, it is helpful to have measured the time they spent talking in relation to how much time other people spent talking. When you have had the discussion with them, either it can help tremendously because they just needed to understand the situation, or it can have no effect at all, since they are unable to shorten a story to the most important parts. I see the latter often with people on the borderline of the autism spectrum, who find it difficult to summarize because all the details seem important to them.

In those cases, my favorite strategy is to agree on a discreet sign I can give them at the retrospective, so they know when they have to end the story and ask for questions instead. Generally, the team will have few if any questions. Often, the **Loudmouth** is happy to have been made aware of the issue, since he or she might be able to apply the lesson in other situations by learning how to summarize a story or at least to recognize when it is appropriate to stop talking.

ONLINE ASPECT

In an online retrospective, this antipattern could easily be solved by muting the **Loudmouth**. That is not a good solution, though, unless it is something you have already planned with the **Loudmouth** as the way to deal with this issue. In some ways, this antipattern is more easily solved in an online retrospective, because it is more accepted to make use of round robins where everybody has to say something in an online setting. You can also put the participants into different breakout rooms so that the **Loudmouth** is contained. Since online retrospectives generally are shorter than offline retrospectives, you have to be very aware of this antipattern and find a solution as fast as possible.

PERSONAL ANECDOTE

For this common antipattern, I have many anecdotes. Let me tell you about two from each end of the spectrum.

One was at a company where I often facilitated retrospectives. The company did not have retrospectives on its own, only with me as an external facilitator. One of the people on the team was a **Loudmouth**. He and I had studied computer science together, so we already knew each other well, and we were both aware that excessive talking was an issue for him and thus for everybody around him.

He was intelligent and funny, and people often enjoyed his monologues, but in a retrospective, where people are forced to stay and time is limited, the entertainment factor was not helpful. It was easy for me to find a solution in this scenario, because I knew this likable **Loudmouth**, we had a relaxed relationship, and I knew that his ego could take almost anything.

In this case, I would make fun of it at the retrospective with comments such as, "Thank you, we have heard from you now; maybe someone else would like to say something" or "Oh, I can always count on you for every detail, but we might need to speed this up a bit. Could someone else explain this as well?" This was, in the situation, funny, and it relieved the tension that arose when he started talking. The tension occurred because people were worried about when, or if, he would stop talking and give someone else a chance to speak.

Of course, I am not asking you to start making fun of the retrospective participants, but in rare cases, this can actually be the most effective and least painful thing to do.

Another example was the first time I facilitated a retrospective for a new company. When I am invited or hired to do a retrospective, I ask beforehand about various things. One of the questions I ask is whether there are any tensions I should be aware of. Another is, "Who likes to hear themselves speak?" because then I can prepare for these challenges by preparing different agendas I can change to on the fly. For example, when someone starts to show aggressiveness, I can change to a silent, reflecting exercise, or I can divide the team into smaller groups to ease the tension. The tension, in my experience, often builds up when everyone is watching.

In this case, though, the person who had invited me happened to be the **Loudmouth**, and he was unaware of his own tendency to take over the entire room by talking for half an hour at a time. My host did not warn me about his own verbosity, of course, so I did not know about it until I started facilitating. Luckily, my opening question was to ask the team members each to explain in two or three words how they felt about the last sprint.

This is a dead giveaway exercise for a **Loudmouth**, of course, since two or three words often turns into 20 to 30 words. This person simply cannot or will not shorten what he or she has to say. In this case, I had some time to change my plan, and I opted for activities with a lot of writing and very little plenum discussion. After the retrospective, I was invited to facilitate another retrospective, and before the next one, I had a chance to talk to this **Loudmouth** and explain the issue. He was receptive to what I had to say, and the **Loudmouth** became less loud.

As one final example, I met a new team recently that boasted not one but two **Loudmouths**. Since I was unaware of this situation until it was too late to react, one of them took over most of the discussion time in the retrospective. Afterwards, the other **Loudmouth** confronted me furiously. She was angry with the other **Loudmouth** and also a little angry with me, because she had been allowed to say almost nothing. But actually, she had taken up the rest of the speaking time herself, and I found it hard not to smile a little bit. It is often the case that what makes people most infuriated with others is the shadow of themselves they see in them. Listening to their frustrations can give you a hint about what they struggle with themselves.

Silent One

. . . in which a team member chooses to be almost completely quiet, and the facilitator applies various tactics to make sure the Silent One is heard

Chapter 19

CONTEXT

On our team, Kim is the **Silent One**. She is not a native Danish speaker, and she is shy. The result is that she is very quiet both at the meetings during the sprints and at the retrospectives. She is always smiling and says that it has been a very good retrospective when there is a retrospective-of-the-retrospective at the end. Still, she does not ask any questions or enter into the discussions.

Sarah does not register Kim's silence as a problem at first, since she has to focus on keeping the retrospective moving forward and getting the expected results. Kim is adept at playing the role of the **Silent One**, and this is what her colleagues have come to expect from her. When Sarah decides that Kim's silence is a problem, she calls Kim out at every retrospective and asks her in the plenum discussions if she has anything to add. She never has.

GENERAL CONTEXT

People are not always in the mood to speak, and some are more talkative than others, but some people are quiet to an extreme. It is nevertheless important to hear what this latter group has to say because a retrospective is a team activity, and everybody on the team should be heard. Otherwise, only those who shout loudest will get their say, and the **Silent One** will rarely be heard.

Often, the **Silent One** is someone who feels less important than the rest of the team—a student intern, a new arrival to the group, a member of a minority of gender, nationality, or work role. I often see testers as the **Silent One**, not necessarily because they are shy by nature but because the team has taught them that developers' opinions are more valuable than those of testers. This attitude is changing now, but it is changing too slowly for my taste.

Antipattern Solution

What happens most often is that the **Silent One** is not detected because the facilitator is busy moving the retrospective forward, keeping the people who are talking focused, and making sure the **Loudmouths** don't dominate the meeting. It takes a pretty experienced facilitator to detect **Silent One**, since they are so amiable most of the time. (If they are not amiable, see the antipattern **Negative One** in Chapter 20.) Nobody really minds that they are silent, since they seem to be happy. And if they have nothing to say, why pressure them?

Consequences

There are many negative consequences to this antipattern. The most important one, I believe, is that not every voice is heard, and the team loses the benefit of shared experiences. The **Silent One** effectively removes him- or herself from the discussions and thus from the decisions. As a quiet listener, the **Silent One** often has had a chance to observe and think about all that has been said, so this person is well worth listening to despite the effort it takes for him or her to speak.[1]

Another negative consequence is that people who already feel only marginally a part of the team can feel even less like part of it if they are kept quiet in a corner, even if it is by their own choice. The team then misses diverse insights. In addition, it is hard to know if the **Silent One** is secretly unhappy or is thinking of leaving the team or the organization.

Symptoms

One of the team members is always quiet. You might not notice it at the beginning, but after a few retrospectives, you start seeing a pattern of behavior—or rather, an *antipattern* of behavior! You find yourself asking this person specific questions or giving him or her extra attention to

---

1. Remember, still waters run deep.

encourage the **Silent One** to contribute to the discussions. You might also notice someone is silent because every time he or she tries to say something, someone else interrupts—in which case you have to look at the **Loudmouth** antipattern (Chapter 18).

REFACTORED SOLUTION

Depending on the situation, this antipattern can be remedied in different ways. One obvious part of the refactored solution is also mentioned in *Agile Retrospectives: Making Good Teams Great* (Larsen & Derby 2006): make every person in the room say something early in the retrospective. Once they have said something, it is often much easier for them to speak again, and it is definitely the case that if they are allowed to be quiet, it is much easier for them to stay quiet. This is also called the *activation phenomenon.*[2] So, always start your retrospectives with everybody saying at least one word.

If I notice a **Silent One** in my retrospective, I start dividing the team into smaller groups when they need to discuss an issue. I start by dividing them in two groups or sometimes into groups of three. If the **Silent One** still seems reluctant to talk, I use the *Think, Pair, Share* approach from my teaching training.

Think, Pair, Share

Think, Pair, Share is a technique in which, instead of asking a question in plenum, you ask a question and let people *think* on their own first. Then, after 30 to 120 seconds, depending on how hard the question was, you let them discuss their answer in *pairs*. After these two rounds, they should have the opportunity to *share* their answers in plenum if they want to.

---

2. In *The Checklist Manifesto* (2011), Gawande describes how important it is to activate people from the start in order to enable them to be active later.

You can also use the *1, 2, 4, All* technique from *The Surprising Power of Liberating Structures* (Lipmanowicz & McCandless 2014) in which there is an additional step between the *pairs* and the *sharing*—the pairs go together in groups of *four*.

For both activities, it is important that everyone has a chance to think and reflect on their own before they start talking to someone else. Some people have an active preference and others have a reflective preference in the way they think, learn, and work. Active thinkers need to talk about issues directly. Reflective thinkers need some time on their own to reflect on the issue before they start talking. If you just ask a question in plenum, they will never be the ones to answer first, and once the plenum discussion has started, they will never have the space to reflect.

To help myself recognize the **Silent One** antipattern, I often write down the names of all the people present in the room. As each person contributes to the discussion without being directly prompted, I add a little dot next to his or her name. If anyone on my list winds up without a dot, I look closely at whether he or she might be a **Silent One**.

When you know that you are in this antipattern and the tricks I've outlined do not work, you might have to talk to the **Silent One** outside the retrospective to see if there is a particular problem at the retrospective and whether you can do anything to help. Otherwise, splitting groups up or having activities with less talking can be the refactored solution to this antipattern.

ONLINE ASPECT

In an online retrospective, it is easier for a **Silent One** to hide, and you might not be aware of the situation right away. When you have acknowledged that you have a **Silent One**, the online setting can work to your advantage. You can make use of round robins, breakout rooms, and activities that involve no talking, just writing, and in which all the team members are muted. The point is not so much to force the **Silent One** to speak as to make a context in which the opinions of the **Silent One** can be heard. Some of the online retrospective tools can help you with this task, because they often include a lot of support for writing and drawing together.

PERSONAL ANECDOTE

In a team where I facilitated retrospectives every other week, I noticed that one of the members was very quiet. I did not know why, and since it was a distributed retrospective and he was at a different site, it was not easy for me to casually talk to him. I wondered, naturally, if he felt that the retrospectives were less useful than he needed them to be, so I decided to try something with the team.

At the next retrospective, they had populated a timeline with Post-it Notes about the events of the previous 2 weeks and how these events had made them feel. Instead of reading them all aloud or allowing people to read their own notes, I decided to make a sort of round robin in which they all had to first choose a positive Post-it Note to discuss.

They could choose their own Post-it Note, one they were interested in discussing because they found it relevant, or one they did not understand. Most people did not choose their own note but instead chose one of the latter two types. We went through this exercise first with all the positive notes, then the question mark notes, and then the negative notes.

Although most notes were chosen because people found them interesting, it was an eye-opener to me how many were chosen because people did not fully understand what was being referred to on the note—either the issue itself or the reason it was viewed as positive or negative. This exercise opened up the **Silent One** in a way I had never seen before. I did not choose him first for any part of the exercise, but once he saw other people choose issues they did not quite understand, he was able to pick those that he did not understand.

I realized that he had been holding back previously because he was afraid to admit that there were things he did not understand. So, in a very nondirect way, we created more trust among the team members, and that was exactly what was needed in this instance of the antipattern.

Negative One

. . . in which one team member's attitude has great negative impact on a retrospective, and the facilitator shields the other team members from the negativity

Chapter 20

CONTEXT

If we look a bit back in time in the story of Sarah's journey on the road to becoming an experienced retrospective facilitator, we are reminded of the very first retrospectives, where Rene was less than impressed with the whole retrospectives business. Rene was late for retrospectives and apologized for it in passive-aggressive ways, such as by saying, "Oh, sorry, I was so busy doing my real job" and "I am glad everybody else has time for games at work."

During the retrospectives, Rene took every opportunity to make ironic or sarcastic comments—always with crossed arms and a smirk on his face—about Sarah's facilitation or the activities she chose. Sarah tried her best to keep Rene engaged in a positive way in the discussions, but she felt she got nowhere, and she could see that it was affecting the rest of the team.

Because Rene was a skilled programmer who enjoyed a high status in the team, everyone looked to him to see how they should respond to things, and his negativity soon spread to the rest of the team.

GENERAL CONTEXT

Negative Ones are very open about how they feel about retrospectives. They ridicule them or make derogatory remarks about a particular retrospective or retrospectives in general. They also try to win others over to their way of thinking in an attempt to end the retrospective or at least make sure there are no more of them in the future. This behavior can cause a lot of worry and stress for the facilitator, since it is very hard to ignore. Most likely, the **Negative One** is also negative in general and perhaps takes pride in being the critical voice.

ANTIPATTERN SOLUTION

Many facilitators choose to ignore the **Negative One** and hope the behavior will pass. Some facilitators become affected by the **Negative One's** attitude and become angry or upset or feel inadequate at facilitating retrospectives. If not dealt with, this negativity can spread like cancer throughout a team.

CONSEQUENCES

A single **Negative One** in a retrospective might not be a huge problem, as long as the negativity does not spread among team members. But if the **Negative One** is popular or is highly esteemed by the team, it might spread to the rest of them. This will make the team members uneasy and the retrospectives unpleasant and less useful because participants will be afraid to take the activities seriously. And when the team fails to take the retrospectives seriously, it is hard to share and thus hard to accomplish the goals. If you are the kind of facilitator who gets angry[1] or upset, the **Negative One** antipattern can have consequences for your mood and your feelings about yourself not only as a facilitator but also as a person.

SYMPTOMS

Negative Ones have their arms crossed, literally or figuratively, or both. They giggle, snort, or whisper negative or "funny" things to other team members. They might also talk in a negative way about retrospectives in the hallways, but you will not always hear about this. Another symptom is how this behavior affects the other people on the team. They might start sharing only superficial events with the rest of the team, or they might not participate in the activities during the retrospective.

REFACTORED SOLUTION

In my experience, the **Negative One** is not negative to be evil or to consciously hurt others.[2] This individual is often happily unaware of the effect

---

1. If you know me, you might find it hard to believe that I have had to work hard with anger-management issues when it comes to retrospectives and teaching. If you don't know me, you probably find it easier to believe that a huge Viking can go beserk from time to time.
2. This is not entirely true, since psychopaths do exist.

of his or her actions or words. Try not to take the behavior personally. I know it is hard, and I still find it almost impossible at times not to take it personally when someone "hates retrospectives" or finds them to be "a glorious waste of time" (see my footnote about going berserk). Think of the **Negative One** as a gift you can use to learn about him or her, about the team, and perhaps about facilitating retrospectives in general. The gift might be wrapped in smelly, ugly paper, but it is still a gift if you choose to look at it that way.

Remember, the Prime Directive holds for you as well. You must do your best to believe that everybody is doing the best they can with the resources available and the knowledge and energy they have at the time.

Now for what you can do about the **Negative One**. Take a moment to reflect on what he or she is saying and when. Is there something that could be triggering the behavior? Is there something this person might be afraid of? Some people are afraid of changes in general, and a lot of people are afraid of a transformation toward agile ways of working because it increases transparency. If everybody knows the status of everything, that could mean everybody knowing that you are stuck with a problem, which can be frightening. Having everybody know when you are not pulling your weight can be even more frightening. Some people fear change in general because they are unsure whether they can work in a new way or can understand the new process.

Take some time to meet with the **Negative One** outside the retrospective, one-on-one, and explain that his or her behavior affects other people and that it affects you as well. Be as honest as you can and want to be.

You can also apply the very effective pattern from *Fearless Change: Patterns for Introducing New Ideas* (Rising & Manns 2005), the Champion Skeptic. In short, this pattern asks you to make the **Negative One** work in your favor (or in favor of the change you are helping to implement, in this case, the use of retrospectives). You tell the **Negative One** that you know he or she is intelligent and knowledgeable about the workplace, the team, and the system and that you have noticed his or her ability to spot critical problems with using retrospectives as a learning tool.

You explain that you need his or her help, thereby turning the **Negative One** into a Champion Skeptic. Retrospectives are an undertaking the team has decided to do (or perhaps have been forced to do), and your new Champion Skeptic can help you figure out what works and what does not work. During the retrospective, you can have a, perhaps secret, agreement whereby he or she signals you if something needs to change or if something happens you should be aware of. Then, together, you can make a plan for the next retrospective.

I have used this refactored solution numerous times, and the Champion Skeptics have become the most brilliant champions for retrospectives not because I have manipulated them but because they feel like part of the movement. This solution only works if you have the respect of the Champion Skeptics, but since they often are negative due to inferiority complexes or lack of acknowledgment, you might gain their respect by acknowledging them. After all, in their mind, if you find them intelligent or accomplished, you must be intelligent yourself.

ONLINE ASPECT

As in offline retrospectives, it is important in online retrospectives to find the causes behind the negativity of the **Negative One**. Once you know what causes the negativity, you can start working with it in the same ways as described previously. If you find yourself suddenly meeting a **Negative One** in an online retrospective, you have to make the most of it in real time. An online retrospective offers ways for you to contain the negativity either by making all activities written and not oral, arranging more voting than free text, and using breakout rooms. The important aim is to not let the negativity spread to the rest of the team.

PERSONAL ANECDOTE

The most **Negative One** I ever had was a woman at a large Danish IT company. It made me wonder at the time if she was so negative toward retrospectives and everything agile because she had been forced to act like "a tough man" to be taken seriously in the IT industry. I often saw this behavior in women: they felt they had to toughen up to the point where any activity other than typing code was dismissed as ridiculous. Whatever her reason, I immediately noticed that she was hard on me and critical of the retrospective. I did two things.

First, I focused on the people who were interested and eager when I went through my introductions to activities and the agenda. More than 20 years of teaching and presenting have taught me that you make yourself happier if you seek eye contact and reactions from the people who are actually interested instead of spending a lot of mental energy on those who would rather sleep or play or work on other tasks. The people who are not interested will not be convinced by words in general anyway. As in everything else in life, you have to decide what problem you want to spend time on at the moment, because you will not have the time or energy for everything. This is described as the pattern Pick Your Battles (Rising & Manns 2005).

Second, before and after each activity, I emphasized why we were doing/had done it and what the expected/realized outcome was. By making the process of a retrospective concrete and obvious, you help the **Negative Ones** to understand that there might be something in it for them. It is always a good idea to debrief the team after each activity, but in this case, I had to make it very clear what we did and why. I also made sure the **Negative One** was part of every discussion by dividing the team into groups of only two people. In that way, she was unable to just look at her phone, as she was inclined to. (Some people believe they can multitask and read their phone while talking with others, but that is not how the brain works,[3] and it is disrespectful to do so, no matter what the circumstances are.)

---

3. Unless it has to do with different parts of the brain—for example, I can whistle along to the Mozart record in the background while writing this chapter.

She was consistently negative during the entire retrospective, and I decided to talk with her afterwards. It turned out she had tried some agile process before, and in her experience, it only slowed her down. We discussed this a bit and agreed to help each other. She would give it an honest try at the next retrospective, and I would make sure the action points the team had decided upon were followed up and that the gain we expected to get from them was visible at the next retrospective.

We both did as promised, and the result was excellent. She informed me about the parts of the retrospective she felt were less useful instead of airing the negative thoughts to everyone present. I explained to her why I had chosen to do these activities and what I expected the team to gain from them. I sometimes changed course on the basis of her comments, and I always made sure that she noticed everything the team gained from the retrospectives. I did this by asking the participants at the start of each retrospective about the status and the effect of the action points (experiments) they had decided on at the last retrospective.

I also asked, after the Gather Data phase, if there had been any new information for anybody, since a big part of the gain from a retrospective is to get the common picture of what has happened since the last retrospective. And once in a while I would go through all the action points from the last 5 to 10 retrospectives to make it obvious that the team had made some changes and that some of these changes were sustainable and effective.

My former **Negative One** became one of my best ambassadors for retrospectives in the company because she was open-minded enough to learn to appreciate their value.

Negative Team

. . . in which the team wants to talk only about the negative things because they think these are the only things they can learn from, and the facilitator shows them that a focus on positive aspects can be equally valuable

Chapter 21

CONTEXT

The team at Titanic Softwære A/S has now been doing retrospectives for some months. Team members have learned that they can meet some of their challenges and solve some of their problems with retrospectives. They dive right into the problems when they meet for the retrospectives. When gathering data for the retrospectives, it is often the negative Post-it Notes they reach for. When Sarah asks why, the answer is that they are here to solve problems, not to relax in a happy hippie place.

GENERAL CONTEXT

You have a team that tends to focus on the negative issues and events during the retrospectives. Perhaps they have a negative mindset, or because they are developers and thus problem solvers, perhaps they are most comfortable when they are actively resolving issues. Or they might think the sole purpose of retrospectives is to discover what is *not* working and fix it. Whatever the reason, a glum atmosphere can permeate the retrospectives if all of the negative aspects of the team's technology, work, and cooperation are emphasized.

It is a human condition to focus on the negative, and we generally have to consistently work with it. This search for "danger" helped our ancestors survive many life-threatening situations, but it can be a burden when we want to focus on and celebrate the good things in life. Bad news sells better than good news.

ANTIPATTERN SOLUTION

When the team in a retrospective wants to focus on the negative in order to solve as many problems as possible, you will often be tempted to follow their lead. You do the retrospectives for their sake, after all, and you want them to get out of it what they think they need to get out of it. Therefore, you rush past the positive data to take action on the negative—the problems and the challenges. The positive notes, if there are any, are left behind, almost unnoticed.

Consequences

It is satisfying, even rewarding, to solve the team's problems and challenges during retrospectives. We must not forget, however, that retrospectives are about making good teams great, and the team can learn also from the methods and processes that already work. Positives sometimes can be made even better, or you can at least make sure they are not forgotten.

Imagine a distributed team that makes polite small talk on Slack because they recognized in a previous retrospective that it helps them feel connected, as a team should. Their retrospectives are typically peppered with "Good morning," "Good bye," and "How are things for you?"—comments that show courtesy and caring for each other.

Now, in the current retrospective, suppose the team discusses how to use Slack, and they agree that they should not be using it for trivial small talk. Suddenly, the polite comments that remind them they are people who care about each other are forgotten, and some people feel they are no longer part of a team. If they are fortunate, perhaps this issue will pop up again at the next retrospective, and they can revisit the purpose of Slack and the benefits of small talk. But this negative blip in the retrospectives would not have occurred in the first place if the team had focused not only on negative but also on positive issues. They might have remembered that small talk on Slack is a deliberate practice for this team because it reinforces respect for and courtesy toward each other.

Symptoms

The symptoms at the retrospective are many negative Post-it Notes, lack of discussion about the positive aspects of the team and its work, and an often negative atmosphere because the team focuses on the things that are not working and that take away their energy. You may sometimes see the facilitator at a retrospective like this trying to brighten things up with jokes and funny stories. But this effort does not work if the baseline is dark. Trust me, I have tried it.

REFACTORED SOLUTION

The aim is to help the team focus on the positive parts of their work. This is sometimes easy, and all you have to do is say, for example, "Today we will spend a bit more time on the positive parts of the past. What is already working for you? What do you like about working this way?" It might help to add that even though we appreciate that we learn more from failure than we do from success, there is value in looking at what already works and perhaps could be even better.

However, with some teams—and you might not know whether yours is one of them before you try this solution—you need to be more persuasive. For example, you can plan a *positivity retrospective* and tell the team that they can talk *only* about positive events and issues.

A third option is to be rough and simply remove all the negative Post-it Notes after gathering the data. You can read an example of this approach in the anecdote section.

The first option skews the focus in the retrospective toward a more positive mindset, while the second two enforce a retrospective focused solely on positive aspects. Whichever of the three options you choose, you should prepare yourself for pushback. Some people simply do not feel a retrospective was productive if it has not solved some problems. In my experience, though, once they have tried including positive issues in their focus and you have followed through by believing in yourself and the process, the team will see the point and feel happy about it. I would make a fully positivity retrospective with no focus on negative things no more than once every 3 to 6 months, depending on how often you have retrospectives with the team.

This refactored solution often leads to a natural inclusion of the positive along with the negative issues at the "normal" retrospectives.

As a side note, I have often seen negative comments subtly phrased as positive comments, such as "It was great that the build worked a year ago" or "I really like when the manager is working from home." Be aware of negatives in disguise, and point them out with a smile the moment you notice them. Make a note of such comments, though, for next time, or even place the Post-it Note in a parking lot for the next retrospective.

ONLINE ASPECT

As with offline retrospectives, the aim is to make the team focus on all the things that are going well and to celebrate and enhance them. The solutions described earlier are equally easy (and hard) in an online retrospective.

PERSONAL ANECDOTE

(Attributed to Simon Hem Pedersen, one of my braver facilitator colleagues.) For a retrospective with a team that historically wanted to focus solely on the negative aspects of the stories, Simon asked them to make a timeline, and he called a break when all the Post-it Notes were on the board. During the break, he removed all but the green, positive Post-it Notes, so when the team came back, the red and yellow notes were gone. They then decided which cluster of green notes to focus on and did a fishbone exercise to find the reasons behind the green notes. In this way, they focused on all that had gone well and developed experiments on how to make those aspects of their work even better.

Despite Simon's initial fear that the team would be disappointed that he had thrown away the negative Post-it Notes, it turned out to be one of the best retrospectives Simon had facilitated with that group.

Lack of Trust

. . . in which the team members do not trust each other enough to share anything of importance in the retrospective, and the facilitator helps them build that trust

Chapter 22

CONTEXT

In the Danish team, Sarah senses a problem. Kim had been working on a task for some time, and when others asked what was happening with the task or if Kim needed any help, the answer was always, "No, I am fine. I don't need any help." The other team members respected this and time passed. When it was time to make use of Kim's work, because other work was depending on it, it turned out that Kim had been stuck for a while but hadn't wanted to admit it. Everybody got very mad, and since then, the atmosphere in the retrospectives had been uncomfortable. Sarah felt she should address the issue in the next retrospective, so she opened with an activity in the Set the Stage phase that would evaluate the level of safety felt among the team members.

Measure Safety Activity

This activity is also called a *temperature reading* and is a way of measuring safety among people in a group. With anonymous votes, the people choose one out of five numbered statements: 5 = "I feel I can share anything with the team"; 4 = "I am not afraid to share my problems with the team if I think that it is valuable for them or me to do so"; 3 = "I can share what is needed to work together"; 2 = "I do not feel I can share if I have any problems or issues"; and 1 = "I will never share anything that means anything to me with the team because I am afraid of the consequences." The point of this exercise is to measure one aspect of the psychological safety in the group to learn what the level of trust among the people is.

As Sarah had feared, none of the team's answers were higher than 3, which showed her there was a lack of safety among the team members: they did not trust each other enough to show vulnerability. She pointed out that this was not an ideal scenario and asked them to be nicer to each other. That did not change the situation, unfortunately, and even mentioning the Prime Directive at the beginning of retrospectives did not make much difference. See Chapter 2, **Prime Directive Ignorance**, for more context.

GENERAL CONTEXT

Sometimes you find yourself facilitating a retrospective where you sense something is not right. The participants are not laughing together, they avoid eye contact and touch (this can also be a sign of being an introvert, so in itself it is not enough to make any judgment), they are slow in writing things to share in the Gather Data phase of the retrospective, and the issues they do write down are either positive/neutral or very shallow (e.g., "the coffee is too strong").

ANTIPATTERN SOLUTION

The easiest and quickest solution is to ask the participants in the retrospective to be honest and to say whatever is on their minds. You might also try to encourage them to speak more openly by offering chocolate or by ensuring that their comments are anonymous. Some facilitators ignore the low level of trust by trying to force people to share secrets with one another.

CONSEQUENCES

If there is a lack of trust among the team members, then nothing of high importance is shared at the retrospective, and the team takes only superficial actions to adapt to the situation. Consequently, the retrospectives become unproductive, and they might be abandoned. The team members become less and less trusting of each other, and when mistakes are made, they are covered up instead of treated as learning opportunities.

SYMPTOMS

Participants in a retrospective do not want to share anything other than positive issues or very shallow negative issues. People look away instead of looking each other in the eyes. You might also see people's feet trying to make them walk out of the room.

Our Talking Feet

An interesting thing about our body language is that, even if we try to hide how we feel by being aware of eye contact and use of hands, our feet might still betray us. The way our feet are pointed gives a clue as to where our attention is, and it might not be on whatever we are looking at. Sometimes, our feet will try to get us out of an uncomfortable situation even if we try our best to pretend that we are perfectly comfortable. This is a anecdotal phenomenon and not supported by research, but I have sometimes noticed it with students giving presentations about an assignment they are unsure of or first-time speakers at conferences: while they are facing the audience and trying to make eye contact with them, their feet start to shuffle toward the door. It is as if the feet are trying to say, "Let us get out of here. I will take over, since you are obviously too stupid to take care of yourself."

REFACTORED SOLUTION

The first thing to do when you suspect a lack of trust is to figure out if it is as bad as you suspect, because you might be making a false assumption. One way to get your answer is to use the safety-measuring activity that Sarah chose. You can find numerous other activities online to help measure trust, but this one is simple and easily done. With activities such as this, you naturally have to consider the *Hawthorne effect*,[1] but doing nothing is worse, in my view. You might get a result that is not 100 percent correct, but in my experience, it will still tell you at least part of the story.

If the result is as you fear, your next challenge is to decide if you are willing and able to work toward increasing the trust level or whether you will work with the team given the existing low level of trust. If you choose to try and increase the trust level, the next thing you have to do is to figure out whether the team is interested in changing it. One team I visited a few times informed me that this was a workplace and any talk about feelings or private life was frowned upon. I did not have a long relationship with

---

1. The alteration of behavior by the subjects of a study due to their awareness of being observed.

that company, since this was the culture encouraged by the management, and I decided that the best thing I could do was to protect myself by leaving. Naturally, I wanted to put all the team members in my pocket and take them with me and take care of them, but that was not possible.

Let's take a step back before we go into the solutions and look at what trust is and why it is considered important among people who work together. (I know you think it is obvious, but indulge me for a moment.)

Let us look at two definitions of trust:

1. Trust is the "confidence that [you] will find what is desired [from another] rather than what is feared" (Deutsch 1977). If you think about it, the duality of hope and fear is exactly the way you can choose to look at your future interactions with other people. If you trust people, you feel confident that you can get what you hope for, whereas if you distrust people, you fear their reaction.

2. Trust is the intersection of a person's hopes and fears (Simpson 2007). Here again is a definition that uses the border between hope and fear to describe trust. It is important to distinguish between these two expected types of outcome in interactions with other people. You can make your own evaluation about whom you trust and distrust by writing down how you expect them to react to different scenarios. For example, will my colleague still do what she promised me to do when she gets busy with her own tasks?

Someone once described trust as a sum of relationship and reliability, and that equation has been my companion for years. It made a lot of sense to me, and in all cases where I wanted to build trust, I used it. I could see how a relationship between people could enhance the level of trust and also how people who could rely on others to do what they had promised or to let them know if they were no longer able to do it could equally enhance trust. Another thing I learned, or became aware of, is that trust is hard to build but easy to destroy.

I used my simple worldview and it was useful to me, but recently I have been made aware of research in trust and better definitions of trust. In McKnight and Chervany (2001), you find the definition of the topology of trust as follows:

- **Benevolence** means caring and being motivated to act in one's interest rather than acting opportunistically.
- **Integrity** means making good faith agreements, telling the truth, and fulfilling promises.
- **Competence** means having the ability or power to do for one what one needs done.
- **Predictability** means trustee actions (good or bad) that are consistent enough to be forecasted in a given situation. Predictability is a characteristic of the trustee that may positively affect willingness to depend on the trustee regardless of other trustee attributes.

In other words, a trustee who is predictable in being benevolent and has the competence to live up to the trustor's interests with integrity is a person worthy of trust. As you can see, my initial equation of reliability and relationship fits well into this topology but is a little coarse.

We can also talk about a *trusting stance*, whereby some people believe they can achieve better outcomes with other people if they assume that those other people are well meaning and reliable. This is likely a personal choice or a strategy people use that colors the way they work with other people. Personally, I had this trusting stance to a degree that made me an easy target for people who wanted to abuse it. Because people have taken advantage of that stance, I have, over the years, developed a less trusting stance, which makes me sad because I would prefer to expect the best from people. In conclusion, if people erode your trust, it can erode your trusting stance. It is a difficult goal to maintain a trusting stance.

And what happens when you have little trust in the people around you? It can trigger the amygdalae,[2] a part of the brain responsible for detecting danger and triggering a very primitive reaction, the fight or flight response, to an interaction or a person. Overactive amygdalae can interpret or categorize even seemingly insignificant events and interactions as dangerous. In the daily life of the office, this results in people who are unwilling to ask each other for help or to share problems and challenges and who start overinterpreting what everybody else says as provocations or criticism. See Chapter 2, **Prime Directive Ignorance**.

In contrast, in a team with high trust among people, team members are more likely to help each other and to ask for help, thereby utilizing everybody's strengths and skill sets. The result, not surprisingly, is higher-quality work and a more harmonious team.

It is no surprise, then, that a study made by Google found that trust, in the form of psychological safety, is the greatest indicator of team performance. As Paul Santanga, head of industry at Google, said after a 2-year study on team performance, "There's no team without trust."

And how do you build trust in a team with retrospectives? My first step is to talk about what trust is, now using the more detailed definition than the equation I used to pull up. The next step is to describe how important trust is and what the consequences of lack of trust can be.

Then I try different activities and techniques. Most are focused on team building, but some also focus on work and interaction ethics. I try to make them laugh together, because laughing together is a very strong way of team building. You can find funny cartoons or quotes, and if you dare, you can make fun of yourself. Self-deprecation can be very efficient to make people trust you and like you, because you show that you are human

---

2. Two almond-shaped clusters of nuclei located deep and medially within the temporal lobes of the brain in complex vertebrates, including humans. They perform a primary role in the processing of memory, decision making, and emotional responses (including fear, anxiety, and aggression). The amygdalae are considered part of the limbic system.

and not without flaws. You can make the team answer nonthreatening questions about themselves as human beings. An activity such as Two Truths and a Lie is always fun, and you can use QuizBreaker[3] in the time between retrospectives.

Two Truths and a Lie

Two Truths and a Lie is an activity often described as a game for teenagers, but it can also be used with much older people as an icebreaker or a way to get to know each other. There are many different ways to play it, but the essence is that everybody writes down three things: two are true, and the last is a lie that potentially could be true. For the more pedantic audience, you might have to stress that it is against the rules to write things such as "I am an ant." When everybody has three things written down, you take turns reading your three things aloud or post them in a chat. It is now up to other people on the team to figure out which of them is the lie. You can learn a lot about each other in this game, but more important, you have some hooks, some pieces of information, you can use to start a conversation. This is also the time to brag, if you feel like it; for example, you can mention that you have run a marathon, climbed a mountain, or have five children yet are still sane. I normally don't play this game to find a winner or to hand out prizes (though some people prefer that sort of competition). I just play it for the laughs.

Help the team learn enough about each other to be able to show integrity and benevolence, but do not forget to make them work on predictability and competence as well. For example, make them aware of their expectations of each other or even devise rules for interacting with and helping each other—for peer review, for when they can interrupt with questions, for how they want to be invited to discussions, and so on. This code of conduct could be recorded in a team charter and define behavior not just for retrospectives but for all interactions.

---

3. QuizBreaker is a fun weekly quiz delivered by email that helps teams get to know one another better in just 2 minutes per week.

Sometimes, though, the culture in the entire organization[4] needs to be changed in order for the culture in the team to change, so it can be a bigger issue than the retrospective can offer help with. The managers can start by appreciating failures and mistakes and showing how we all can learn from them. They can follow up by celebrating the learning instead of threatening with ridicule or even firing employees. They can promote or otherwise appreciate the people who are trustworthy and good to work with as much as they appreciate the high-performing people. Unfortunately, an organization's culture is usually out of the hands of the retrospective facilitator.

ONLINE ASPECT

For an online retrospective, it is easier to make data gathering and voting anonymous, which is the default setting in some online tools. So, the quick fix is easier in an online retrospective, but the long-term goal of building trust among the team members demands a more focused effort when people are distributed.

PERSONAL ANECDOTE

It is not a bad idea to ask a team what their level of trust is, but be aware that if you ask them, you should be prepared for the answer. Any answer. This is rather like asking your small child, "Do you want to go to sleep now?" or "Can you stop kicking the lady?" If there is only one acceptable answer, you should not ask the question. So, if you ask them, you need to consider in advance what your action will be if the trust is very low.

I will share two anecdotes in which there turned out to be a lack of trust.

---

4. Watch the video by Simon Sinek on Performance versus Trust (https://www.youtube.com/watch?v=YPDmNaEG8v4).

At one company, I was hired as a consultant to facilitate a retrospective. I had asked the sponsor if there was anything I should know about the team. He said that people were a bit afraid of sharing problems with one another. This was great information to get from him, because even if I had been able to detect the low trust level myself at the retrospective, it was more useful to know about it ahead of time. I was now able to plan with it in mind.

I decided to start the retrospective with an intensive presentation of the Prime Directive and a safety measurement activity, as described previously. The team members were very quiet when I talked about the Prime Directive. They did not even move a single facial muscle. This was a bad sign. Also, the results of the trust activity were nonoptimal. One person had complete distrust; the rest had little or barely enough trust. So, the sponsor was right but probably not aware of how big the challenge was with the team. I seriously considered ending the retrospective right there and then.

There is not much value in a session where you are supposed to share information, experiences, and feelings if no one feels safe enough to share anything of importance. I decided to label the feelings instead, as described in *Never Split the Difference: Negotiating as If Your Life Depended on It* (Voss & Raz 2017), by putting a name to the feelings I believed existed in the room. I said something along the lines of "It seems you are afraid to share information, particularly negative informa-tion, with each other. I suppose you are afraid that you will face anger, ridicule, or laughter. I sense that you have some experiences and feelings you need to share, though."

I could see from their body language that I had hit a nerve. They offered me eye contact, more and more as I spoke, perhaps because they felt seen by me. I explained that in this retrospective, it was my responsibility to ensure that everyone could say anything without worry, and I would do my utmost to live up to that responsibility. I paused to give the team time to think about what I had said. Then I asked them if they wanted the retrospective to go on and if they would try to share as much information as they could. There was a murmur, some nodding, and they started gathering data on Post-it Notes.

Had this been an online retrospective, we would have had the luxury of anonymity. That can still be done in real life, but it takes more preparation, since you have to ask the questions beforehand. Had I known that the situation was as dire as it was, I might have done that. Throughout the retrospective, I talked a lot about learning from mistakes, getting to know each other, and asking for and offering help, and I tried my best to make them laugh together. I talked to the sponsor afterwards and said that this team needed some help. He wanted me to go into specifics about who had said what, but I would not break that confidence. Instead, I made him understand that this team could run into some serious issues if they did not start communicating more freely and sharing failures.

Another company was one I actually worked for, so I knew everyone, and I had been able to follow them closely. I knew trust was low, and I had an idea of how to work with it. There were some rumors and misunderstandings in and around the team. I started a retrospective in exactly the same way as in the previous anecdote: with the Prime Directive and a safety evaluation. As expected, trust levels were low, and not many team members felt safe enough to speak up.

In this team, the distribution was a bit different, though, since two of the team members felt completely safe to share everything with the team, and even if the vote was done anonymously, it was not difficult to guess who these two were. Unfortunately, these people were also extroverted bullies, whereas the rest of the team were introverts. Not a good combination.

I tried working on ground rules with the team during this and the following two retrospectives, asking how they wanted to work together and how they preferred to communicate. I tried asking them how previous great teams had been for them and what ways of working had characterized those teams. I made them think back to failures that turned out to be good lessons. I made them think of success stories in the team. I did a futurespective with them to help them share what they feared and what they hoped for. Following my own advice about working with **Loudmouths** and **Negative Ones,** I talked with the bullies outside the retrospectives about their behavior and the consequences of it. But all was in vain.

I was unable to change the dynamics of the team or the way the bullies behaved, and people became unwilling to work in the team. When the first person fled the team, I decided to speak to management about it. I had to name names, I had to point out the bullies, and I had to propose an action.

My proposal was to remove at least one person from this team, either to another team or completely out of the organization. Not everything can be solved in a retrospective setting, and one of the bullies was removed from the team under a lot of protest, but it enabled the remaining bully to change his ways, and slowly the team regained the trust they needed to work together in a fulfilling way.

Different Cultures

. . . in which the assumptions the facilitator or the team members bring from their own culture are preventing them from seeing how the retrospective is experienced by others, and the facilitator finds ways to make them more aligned

Chapter 23

CONTEXT

Sarah has now earned a reputation at Titanic Softwære A/S for being an effective facilitator. She receives invitations to facilitate other teams, and she is happy about that. Not only does she learn a lot more about facilitation because she uses different approaches for different teams and people, but she also learns a lot about what the other teams are working on and struggling with. Some of the things they decide to try out she can even bring back to her own team to discuss.

The company has recently decided to outsource some development. It has hired a new group of developers, Coders by the Hour (CbtH), from a company that has a different corporate culture than that of Titanic Softwære A/S. The employees at CbtH don't want to speak up about problems. They try to hide them and solve them, because problems imply someone is guilty, and at CbtH, you might get fired if you are guilty of something.

Sarah is asked to facilitate a retrospective with this team together with the people they are collaborating with at Titanic Softwære A/S.

GENERAL CONTEXT

Sometimes, people are afraid to speak up. They worry that if they voice their opinion, the consequences will be harsh. The level of trust varies widely from company to company and country to country. Not all companies have ways of sharing feelings and thoughts anonymously, and openly sharing concerns and issues can be frightening.

ANTIPATTERN SOLUTION

It is easy to forget to consider the possibility of new problems with a different culture and instead to follow the usual procedure. The effect is that people clam up. They choose not to share what they have experienced, and as a result, the retrospective suffers because the important issues are not discussed.

CONSEQUENCES

If people are afraid to speak up during a retrospective, issues will be addressed only superficially. It is not easy to build the trust needed to talk about sensitive topics. As a consequence, people will avoid emotional topics and the important issues will remain unresolved. The retrospectives will start feeling—and being—a waste of time.

SYMPTOMS

You experience behavior in the retrospective that you do not understand, and you do not initially know how to deal with it because it stems from a culture and value system that are different from yours. In my examples, the retrospective participants will not share difficult concerns because in their corporate culture, employees do not voice negativity for fear of being fired. The hierarchy in their company works in a different way from what you are used to.

REFACTORED SOLUTION

There are several ways to deal with a lack of trust; some of them are short-term solutions, and others long term (see Chapter 22, **Lack of Trust**). One quick fix that I sometimes use to deal with lack of trust, at least with distributed retrospectives, is to use Google Docs in anonymous mode. You share the document using the "Anyone with the link" option, and the team members access the document through the link *without* logging in to their Google account. Then they show up with aliases such as Anonymous Mouse, Anonymous Manatee, and other anonymous animals (Figure 23.1). People do not know which animal names they are assigned, and they certainly do not know who the others are.

Figure 23.1 Anonymous participants in a Google Drawings document

You may sometimes find yourself working with a team whose fear of consequences is as big and as realistic as actually getting fired in a country where jobs are scarce and people have no social safety net. But, on a smaller scale, lesser fears can feel similar and have the same consequences for openness in the retrospective. Fear of being laughed at and fear of losing face, respect, or promotion opportunities similarly affect openness and employees' willingness to share.

For a more long-term solution, I prefer to work on building trust, as I dealt with in depth in the **Lack of Trust** antipattern (Chapter 22). To build trust, you first have to be aware that there is a problem, which can be done by simply asking the team about the level of safety they feel. There are a lot of activities to evaluate the amount of trust among team members.

Asking about trust immediately causes people to reflect on how they feel, and the question also has the more relational effect of helping people see how others feel about being at that retrospective/meeting with them. Of course, some people's fear of sharing exists on a personal level, and no matter how nice people are to them, they are reluctant to share anything. Lack of trust also may be relational, and making a change in the way the team communicates can increase the level of trust.

More than anything, trying to understand the culture that influences a team, be it a national or an organizational culture, is really helpful when preparing retrospectives.

ONLINE ASPECT

The context of this antipattern is an online retrospective, but the challenge of different cultures can be present in offline retrospectives as well. If the difference manifests itself in silence from half the team members, you can make a quick fix and try to make most of the participation anonymous. This might work, but it is not really a long-term solution, and if you can meet the people in real life, you have more options than in an online setting. You might be able to talk privately with the individuals, before or after the retrospectives, and learn about their culture by asking them about events they have experienced in the workplace, good and bad. If they open up to you, then you might be able to have a retrospective about the difference in culture and together try to come up with experiments to try out at the next retrospective. For example, have a *futurespective* to make the issue a bit more abstract, or set more time aside for small talk at the beginning of the retrospective.

PERSONAL ANECDOTE

In one particular setting, we had a team from India working with our Danish team, and we had retrospectives set up with them every 2 weeks. I was told the people from India never shared anything negative or anything that could be interpreted as such. Perhaps they were afraid of getting fired, or perhaps it was not part of their culture to voice negativity—I did not know.

I set up a Google Doc, used the "Anyone with the link" option, and invited them to the retrospective by emailing the link. The team from India immediately understood how it worked, that they had no idea who was writing what, and more important, that no one else knew what they were writing. Once they got started in the Gather Data part of the retrospective, it was hard for them to stop. They had a lot of issues with the system, with the code quality, with the meetings, and with the communication in general. This outpouring was great because it gave us the opportunity to address issues—important, game-changing issues—that we had never addressed before.

But the best part was when the teams brainstormed for experiments to try out. The Indian team had bold ideas for experiments, things they had never suggested before. They laughed and made jokes at the end of the retrospective, which I saw as a good sign that they felt both heard and relieved to be able to share without fear of consequence.

Dead Silence

... in which the team members are completely silent, often in an online retrospective, and the facilitator uses various tactics to hear their opinions despite their reluctance to participate

Chapter 24

CONTEXT

Sarah felt she had tried everything to encourage her team to talk. She had even shared stories about her vacation in the hope that the team would start talking, if not about the way they work, then at least about *something*.

It was the third online retrospective with the distributed team, and Sarah felt powerless. Dead silence reigned in the virtual room they used, and nothing she said could get more than a small background hum as response. Most of the team members' sound and video were turned off. The ones on video were looking at their screens and tapping on devices, but nothing appeared in the shared document, so Sarah knew they were busy with tasks unrelated to the retrospective.[1]

In her desperation for the team to participate, Sarah even talked the retrospective down, making comments such as "I know this is not your favorite way of spending your time" and "I realize you find these retrospectives useless and boring." The only reaction she got was subdued laughter.

Sarah ended the retrospective early, since there was nothing to discuss.

GENERAL CONTEXT

Often in a distributed retrospective, some people are silent. That can work if you make sure they are heard from time to time. But sometimes the whole team becomes silent, and in contrast to a real-life retrospective, it is not possible to walk toward people or urge them to talk by making eye contact. It is easy to be silent online, but it is much harder in real life, where the silence can get a little uncomfortable.

In online retrospectives, you can have a team placed in a number of different locations, each with their more or less efficient online connection

---

1. As an online facilitator, you get an almost uncanny sense for whether people are focusing on the retrospective or paying attention to something else.

being dead silent. Perhaps they are afraid of speaking up; perhaps they are working or writing emails or scrolling through Facebook; perhaps they have nothing to say about the subject they are supposed to be discussing.

While it seems odd that the team has nothing to say about a topic that they themselves chose as relevant, it nevertheless happens. Perhaps they feel they should discuss the test strategy/process/meeting agendas at this retrospective, because they know they are not satisfied with them. But for some reason, be it shyness, introvertedness, or a feeling of being powerless, the discussion does not happen.

It could also be caused by a mix of politeness and the constraints of online collaboration. Even with the camera on, it can be hard for people to tell who is about to say something, and not wishing to interrupt, they stay silent. Another reason some people stay quiet is that they think what other people have to say is more important than their own input.

ANTIPATTERN SOLUTION

The antipattern solution is to allow people to remain quiet. The thinking is that if they have nothing to say, they should not be forced to say anything.

CONSEQUENCES

In the short run, the consequence is that the discussions at the retrospective are based on very few people's voices, so some causes, stories, or worries are not voiced. In the long run, the consequence is that the retrospectives become fruitless and will be canceled. One by one, individuals will stop making time for them until, in the end, the entire team has lost interest.

SYMPTOMS

The obvious symptom is silence at the retrospective. Blacked-out video and muted sound are also symptoms, although some people mute their sound for other reasons, such as to spare the meeting unwanted background noises—coughing, children interrupting, static, and so on. The less obvious symptom is that people start declining the invitations to retrospectives.

REFACTORED SOLUTION

There are a lot of tips and tricks in the refactored solution because the choice of what to do is very much based on the cause of the silence.

As a general rule, remember to hear everybody's voices at the beginning of a retrospective. If people are "allowed" to be quiet in the beginning, it is much easier for them to stay quiet during the retrospective. This is known as the *activation phenomenon* (Gawande 2011).

Use a round robin from time to time based on the same question for everybody—for example, "How did you experience this incident?" or "To what degree do you think this experiment was successful in the last sprint?" A round robin allows everybody to feel that hearing from them is important even if that is not their usual expectation. I have often heard some surprising insights from quiet people in a round like this. Keep the question simple, at least for the first time, but don't use a yes/no question. Do not make anyone uncomfortable, at least not on purpose.

If you think the silence is caused by shyness, divide the team into smaller groups and let them talk on another channel. If this is not enough to make all of them talk, keep dividing them until they are in pairs. Make sure you are able to listen in on them from time to time.

Sometimes, people are quiet simply because of the sheer number of people in the "room." In some online collaboration tools, you can use breakout rooms. If that is not possible in the tool you use, you can have the team use another tool to talk in smaller groups or even start multiple meetings in the same tool. If you send them out to a place where you cannot reach them, you have to make sure that you have a way to "find them" and tell them to come back if they don't come back at the time you specified.

If the cause of their silence is that they do not feel safe within the team, this probably needs to be dealt with offline or in a separate meeting. If you notice that someone is silent all the time, consider taking up the issue with him or her outside the meeting, perhaps in a phone call or a one-on-one chat. Ask if he or she knows how important it is to hear from *everybody* on the team. Avoid asking, "Why are you so quiet?" The latter can come across as an attack, while the former encourages the person to reflect on his or her role and responsibility within the team.

Perhaps the team simply finds it difficult to talk about things and share experiences and ideas. In this case, you can try to tighten up the agenda for the retrospective. As an example, instead of allowing them 10 minutes to think individually about events from the last sprint, give them only 3 minutes, and then have each person share it with one other person. This gives them a chance to think of something before they have to share it, so the people who need to reflect before talking can be allowed that time. It also gives the people who need confirmation before they share in plenum a chance to test the waters with at least one other person. Of course, in this way, you might not be able to address as many issues as you wanted to (and they wanted to). But focusing on a few subjects and actually having an informed conversation about them is better than touching superficially on 25 topics.

As you can see, there are many ways of solving this problem, and most of them depend on why the team is silent. Finding the cause(s) behind something is important in all aspects of life in general and in retrospectives in particular.

ONLINE ASPECT

The context of this antipattern is an online retrospective, but as mentioned in the Context section, this can happen in offline retrospectives as well. When you are together with people in real life, it is easier to get them to talk, because you can use your body language to encourage them. Another difference is that, in an online retrospective, the disengaged team members are often doing other things, such as reading emails or looking at the news. It is obvious if they do that when you are in the same room, and it is easier to set ground rules about this (e.g., no phones or computers in the retrospectives). It is important to find the cause of this. Is it because they are shy, or is it because they are disengaged, and if they are disengaged, why are they disengaged? When you know more, you might find that you are in another antipattern, such as **Negative Team** (Chapter 21), **Curious Manager** (Chapter 15), **Lack of Trust** (Chapter 22), or perhaps the room is full of **Silent One** (Chapter 19).

PERSONAL ANECDOTE

Some years ago, and fortunately not in the beginning of my facilitation journey, I was asked to facilitate a retrospective for a team that was completely distributed. The team members had never met. They were all from a "strong, silent" culture. None of them had been in a retrospective before. They had been hired for their programming skills, not their communication skills.

I started the retrospective, as always, by asking them all to answer one question. In this case, the question was a simple one about their role in the team. Most answers were short and to the point, which was all I asked for, so that was fine by me.

In the next part of the retrospective, I explained the agenda for them and what I expected them to get out of it. I also explained how to use the Google Drawings document we would be sharing. In retrospect, I think

I could have asked them what their expectations were from the retrospective, but I didn't.

We entered the Gather Data phase in a very quiet way, and I asked them to fill out online sticky notes in a Google Drawing with incidents that had happened in the last 2 weeks and place them on the Google board. Between the eight people present, they wrote 10 notes. I suspected some of the team members had not written any.

Normally, the next part is to go through all the notes and divide them into groups. There was no need for that with so few notes. I read them all aloud to make sure that everyone knew what each note said. Later, I thought that it might have been a good idea for them to read the ones they posted themselves and hear why they chose these incidents to share. But I didn't.

For each of the notes, I paused a little and asked for questions, comments, or reflections in order to Generate Insights. Each time, there was silence.

Since I knew that Generate Insights is a crucial part of a retrospective, I was determined to make the team reflect or at least discuss some of the items. I appreciated that they had chosen to share some things, but I wanted them to learn more from them. One of the incidents was that the planned release had been canceled. I thought that this would be interesting for them to talk about. I drew a fish skeleton and decided to make a "forced" fishbone cause analysis. Normally, I would ask team members to volunteer in writing notes for the possible causes they could think of and place them on the fishbone.

In this case, though, I decided to hold a round robin so that I could force them to speak up. I went through the list of people in the retrospective and made them all give a possible cause for the canceled release, and I wrote it on the fishbone. For the first 5 minutes, people did a lot of sighing, and some said they would like to pass. But after that, the round robin started to work. The result was a number of possible causes for the canceled release, technical as well as organizational. Now we had somewhere to start a discussion that mattered to them, and during the rest of the retrospective, it was easier for me to get them to speak up, especially since I knew my challenge now and was able to compensate by never asking a question in plenum but rather giving them time to reflect and then asking them one at a time.

When I decided to do what I did in this setting, it was because I guessed that the reason for the silence was culture, not shyness or fear. I could have been wrong, of course, and then I would have had to change my approach or even end the retrospective and talk to them all offline in order to know what to do in the future.

Conclusion

The conclusion is simple: Avoid my mistakes. Be aware of these traps in retrospectives, and be mindful they do not make you fall.

That being said, you probably need to make your own mistakes to learn. So perhaps the conclusion of this book is that you will find yourself in most of these antipattern solutions at some point in time, but at least you will know that you are not alone. Somewhere in Denmark, Aino is still feeling down about that particular time she introduced that particular antipattern herself. Perhaps it can help you to know that whatever situation you find yourself in while facilitating retrospectives, I have most likely been in the same. It was difficult for me to write this book because I had to remind myself of all the situations where something went wrong. It was a painful path to take, but I also learned from this overview.

Know also that I have always tried to learn from my mistakes and to improve my facilitation skills next time—even if only a little bit, just enough for me to feel better again.

So, go out and start, or continue, your facilitation journey. Make mistakes, celebrate them or feel bad about them, depending on your nature, but try to reflect on them and learn from them. Some things I have learned from failed retrospectives have been helpful not just within the realm of retrospectives but also outside. For instance, I learned that negativity aimed at me did not necessarily start with me; often, it came from somewhere else. And that people who are quiet often have something important to say if you allow them. I find myself trying to facilitate everything in my life, and I get away with it because facilitation is not about being in power or manipulating others—it is about helping everyone be heard and making sure that people feel as good as they can in every cooperative setting.

But sometimes I am just in an utterly bad mood, and I don't care about other people. And sometimes that is OK.

REFERENCES

Adkins, Lyssa. 2010. *Coaching Agile Teams*. Indianapolis: Pearson.

Alexander, Christopher, Jacobson, Max, Ishikawa, Sara, & Silverstein, Murray. 1977. *A Pattern Language: Towns, Buildings, Construction*. New York: Oxford University Press.

Allspaw, John. 2014. "The Infinite Hows: An Argument against the Five Whys and an Alternative Approach You Can Apply." O'Reilly Radar. https://www.oreilly.com/ideas/the-infinite-hows

Beck, Kent. (1993) 1999. "A Short Introduction to Pattern Language." In *Kent Beck's Guide to Better Smalltalk: A Sorted Collection*. New York: Cambridge University Press, 137–144.

Bens, Ingrid. 2005. *Facilitating with Ease!: Core Skills for Facilitators, Team Leaders and Members, Managers, Consultants, and Trainers*. Hoboken, NJ: Wiley.

Bergin, Joseph, & Eckstein, Jutta. 2012. *Pedagogical Patterns: Advice for Educators*. Pleasantville, NY: Joseph Bergin Software Tools.

Brown, William J., Walveau, Rahael C., McGormick, Hays W., & Mowbray, Thomas. 1998. *Antipatterns: Refactoring Software, Architectures, and Projects in Crisis*. Hoboken, NJ: Wiley.

Caspersen, Michael E. 2007. "Educating Novices in the Skills of Programming" (PhD dissertation). University of Aarhus, Denmark.

Coplien, James O., & Harrison, Neil B. 2005. *Organizational Patterns of Agile Software Development*. Upper Saddle River, NJ: Prentice Hall.

Cornils, Aino. 2001. "Patterns in Software Development" (PhD thesis). University of Aarhus, Denmark.

De Bono, Edward. 1999. *Six Thinking Hats*. New York: Back Bay Books.

Deutsch, Morton. 1977. *The Resolution of Conflict: Constructive and Destructive Processes*. Carl Hovland Memorial Lectures Series. New Haven, CT: Yale University Press.

Eckstein, Jutta. 2019. "Retrospectives for Organizational Change: An Agile Approach." EPUB: ISBN 978-3-947991-00-6

Fowler, Martin, & Lewis, James. 2014. "Microservices." https://martinfowler.com/articles/microservices.html

Gamma, Erich, Helm, Richard, Johnson, Ralph, & Vlissides, John M. 1995. *Design Patterns: Elements of Reusable Object-Oriented Software*. Reading, MA: Addison-Wesley.

Gawande, Atul. 2011. *The Checklist Manifesto: How to Get Things Right*. New York: Picador.

Gonçalves, Luís. 2019. "9 Deadly Agile Retrospectives Antipatterns Every ScrumMaster Must Avoid." https://luis-goncalves.com/agile-retrospectives-antipatterns

Gonçalves, Luís, & Linders, Ben. 2014. *Getting Value Out of Agile Retrospectives: A Toolbox of Retrospective Exercises*. InfoQ Enterprise Software Development Series. Morrisville, NC: Lulu.com.

Kahneman, Daniel. 2013. *Thinking Fast and Slow*. New York: Farrar, Straus and Giroux.

Kaner, Sam. 2007. *Facilitator's Guide to Participatory Decision-Making*. Hoboken, NJ: Wiley.

Kerth, Norman L. 2001. *Project Retrospectives: A Handbook for Team Reviews*. New York: Dorset House.

Kerr, Dave. 2018. "The Death of Microservice Madness in 2018." https://dwmkerr.com/the-death-of-microservice-madness-in-2018

Kurtz, C., & Snowden, D. 2003. "The New Dynamics of Strategy: Sense-Making in a Complex and Complicated World." *IBM Systems Journal*, 42(3): 462–483. https://doi.org/10.1147/sj.423.0462

Larsen, Diana, & Derby, Esther. 2006. *Agile Retrospectives: Making Good Teams Great*. Dallas: Pragmatic Bookshelf.

Lipmanowicz, Henri, & McCandless, Keith. 2014. *The Surprising Power of Liberating Structures: Simple Rules to Unleash a Culture of Innovation*. Seattle: Liberating Structures Press.

McKnight, D. H., & Chervany, N. L. 2001. "Trust and Distrust Definitions: One Bite at a Time." In R. Falcone, M. Singh, and Y.-H. Tan (Eds.), *Trust in Cyber-Societies: Integrating the Human and Artificial Perspectives*. Berlin: Springer-Verlag, 27–54.

Pease, Allan, & Pease, Barbara. 2004. *The Definitive Book of Body Language*. London: Orion Books.

Pelrine, J. 2011. "On Understanding Software Agility—A Social Complexity Point of View." *E:CO*, 13(1–2): 26–37.

Rising, Linda, & Manns, Mary Lynn. 2005. *Fearless Change: Patterns for Introducing New Ideas*. Boston: Addison-Wesley.

Ross, Lee. 1977. "The Intuitive Psychologist and His Shortcomings: Distortions in the Attribution Process." *Advances in Experimental Social Psychology*, 10, 173–200. https://doi.org/10.1016/S0065-2601(08)60357-3

Ryan, Kathleen D., & Oestreich, Daniel K. 1998. *Driving Fear out of the Workplace: Creating the High-Trust, High-Performance Organization*. San Francisco: Jossey-Bass.

Schwarz, Robert M. 2002. *The Skilled Facilitator*. Hoboken, NJ: Wiley.

Simpson, J. A. 2007. "Psychological Foundations of Trust." *Current Directions in Psychological Science*, 16(5): 264–268.

Snowden, Dave. 2015. "Description Not Evaluation" (blog post). http://www.cognitive-edge.com/blog/description-not-evaluation/

Tabaka, Jean. 2006. *Collaboration Explained: Facilitation Skills for Software Project Leaders*. Upper Saddle River, NJ: Addison-Wesley.

Tung, Portia. 2019. "The School of Play." https://theschoolofplay.wordpress.com/who-we-are/

Voss, Chris, & Raz, Tahl. 2017. *Never Split the Difference: Negotiating as If Your Life Depended on It*. London: Random House.

Wolpers, Stefan. 2017. "21 Sprint Retrospective Anti-Patterns." https://dzone.com/articles/21-sprint-retrospective-anti-patterns

Womack, James P., Jones, Daniel T., & Roos, Daniel. 1990. *The Machine That Changed the World: The Story of Lean Production*. New York: Macmillan.

INDEX

CREDITS

Cover: Illustration by Nikola Korać; courtesy of Aino Corry.

Pages xxxi, 16, 18: "Regardless of what we discover, . . . and the situation at hand." Norman Kerth, Project Retrospectives: A Handbook for Team Reviews (Addison-Wesley, 2013).

Pages 3, 5, 17, 27, 37, 47, 55, 63, 75, 83, 91, 99, 107, 115, 125, 131, 137, 149, 157, 167, 175, 183, 189, 203: Illustrations by Nikola Korać; courtesy of Aino Corry.

Page 31: "God, grant me the serenity to accept the things I cannot change, courage to change the things I can, and wisdom to know the difference." Source: Reinhold Niebuhr, American Theologian (1940s).

Page 126, Figure 14.1: Photo courtesy of Aino Corry.

Page 143: "Having heard that he had . . . and our friendship continued to his death." Benjamin Franklin, The Autobiography of Benjamin Franklin (Houghton, Mifflin and Company, 1888).

Page 145: Photo courtesy of Aino Corry.

Page 206, Fig 23.1: Google Drawing Document, copyright Google.

Agile Development
Books, eBooks & Video

Whether are you a programmer, developer, or project manager InformIT has the most comprehensive collection of agile books, eBooks, and video training from the top thought leaders.

- Introductions & General Scrum Guides
- Culture, Leadership & Teams
- Development Practices
- Enterprise
- Product & Project Management
- Testing
- Requirements
- Video Short Courses

Visit **informit.com/agilecenter** to read sample chapters, shop, and watch video lessons from featured products.

Register Your Product at informit.com/register

Access additional benefits and **save 35%** on your next purchase

- Automatically receive a coupon for 35% off your next purchase, valid for 30 days. Look for your code in your InformIT cart or the Manage Codes section of your account page.
- Download available product updates.
- Access bonus material if available.*
- Check the box to hear from us and receive exclusive offers on new editions and related products.

Registration benefits vary by product. Benefits will be listed on your account page under Registered Products.

InformIT.com—The Trusted Technology Learning Source

InformIT is the online home of information technology brands at Pearson, the world's foremost education company. At InformIT.com, you can:

- Shop our books, eBooks, software, and video training
- Take advantage of our special offers and promotions (informit.com/promotions)
- Sign up for special offers and content newsletter (informit.com/newsletters)
- Access thousands of free chapters and video lessons

Connect with InformIT—Visit informit.com/community

Addison-Wesley • Adobe Press • Cisco Press • Microsoft Press • Pearson IT Certification • Que • Sams • Peachpit Press

 Pearson